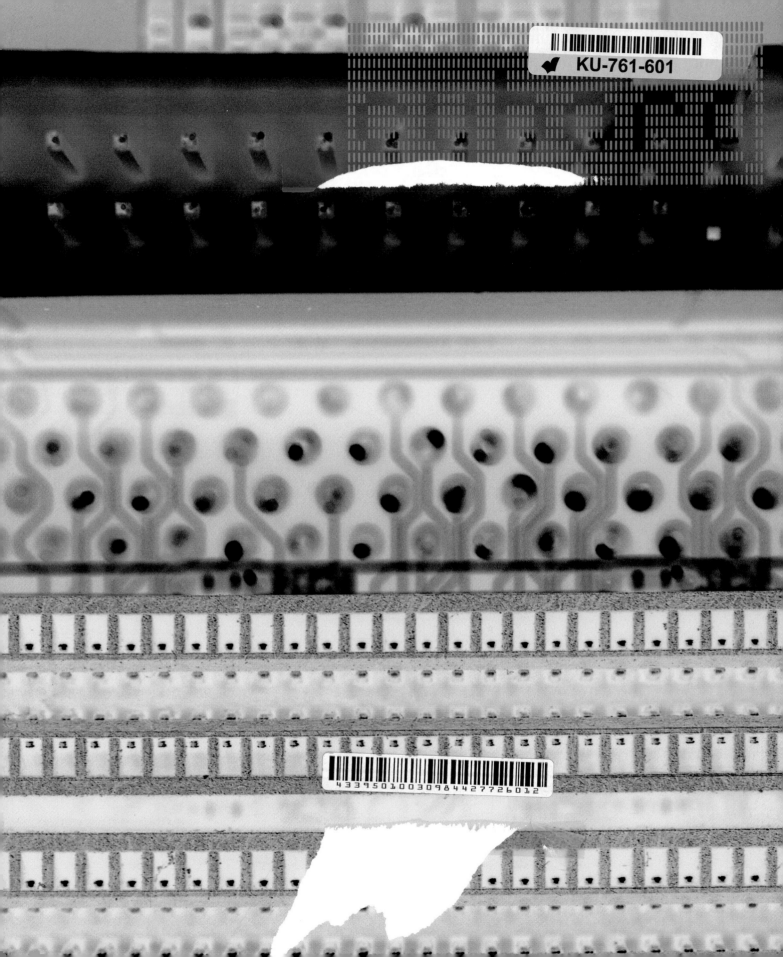

marie o'mahony

# cyborg
## the man-machine

Thames & Hudson

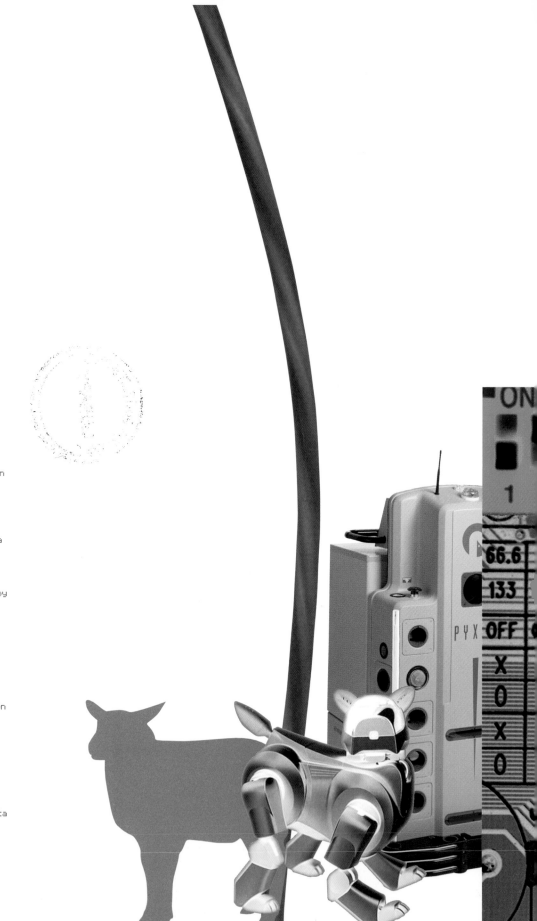

Design Alexander Boxill

First published in the United Kingdom in 2002 by
Thames & Hudson Ltd
181A  High Holborn
London WC1V 7QX

www.thamesandhudson.com

© 2002 Marie O'Mahony

British Library Cataloguing-in-Publication Data
A catalogue record for this book is available
from the British Library

ISBN 0-500-28381-8

Printed and bound in Hong Kong by
C & C Offset Printing Co. Limited

# contents

Human beings have developed very slowly through evolution. During this process, changes have taken place in response to the environment and other physical conditions. The cyborg, part-human and part-machine, could be viewed as a step change in our evolution. But to what extent is the cyborg a progressive and positive change?

01

01 + 02 ●●
The Jacobin Pigeon (shown left) and
Pouter Pigeon (opposite) are both
examples of varieties produced by
artificial selection. Charles Darwin
mentions both in his great work *On
the Origin of the Species by Means
of Natural Selection*: 'The pouter has
a much elongated body, wings and legs;
and its enormously developed crop,
which it glories in inflating, may well
excite astonishment and even laughter....
The Jacobin has the feathers so much
reversed along the back of the neck
that they form a hood.... The trumpeter
and laughter, as their names express,
utter a very different coo from the
other breeds....'

Charles Darwin was a pigeon-fancier and
corresponded with other skilled breeders.
In his great work of 1859 *On the Origin of the
Species by means of Natural Selection*, Darwin
quotes a fellow pigeon-breeder, Sir John
Sebright, who claimed that 'he would produce
any given feather in three years, but it would
take him six years to obtain a head and beak'.
The bizarre results of such artificial selection
are also illustrated in C. Barry's 'A treatise on
domestic pigeons', an eighteenth-century
manuscript, which shows the extraordinarily
diverse pigeons created by breeders. Yet when
Darwin first visited the Galapagos Islands in
1835, he paid little attention to the finches.
It was only later that he concluded that the birds
were in fact an excellent example of evolution
at work. On each island, the finches had beaks
adapted slightly differently to cope with the
different food available. They displayed what
biologists call adaptive radiation, which is
the evolution of a founding population into
an array of different species, each adapted to
its own ecological niche. Darwin was perhaps
better able to perceive the changes to pigeons
that took place through human intervention
because they took place so much faster and
the results were so much more flamboyant.

3

03

Two figures by the Florentine artist
Braccelli, from his *Bizzarie di varie
figure* (Oddities of Various Figures),
Livorno, 1624. The intention of
the series was to puzzle, amuse, and
entertain. The two anthropomorphic
figures are created using household
and garden tools – the scissors and
spade, for example. Technology now
allows the contemporary robotics
scientist to bestow humanoid
attributes on the machine, so much
so that one day it may be difficult
to separate the two.

Selective breeding enables us to steal a march on nature's rather slow process of evolution as far as animals are concerned, but the idea of the selective breeding of humans, advocated in early eugenics, has been discredited since the mid-twentieth century as morally repugnant, whether positive by selecting for 'good' qualities, or negative by eliminating or sterilizing those with 'bad' genes. However, it seems that new scientific advances, particularly in genetic engineering, will soon make it possible to increase the pace of human evolution. The alteration of humans through chemicals and genetic modification is an area that causes the deepest fears, both among scientists and the general public. Unlike mechanical or electronic modifications, these changes are invisible to the eye, and irreversible once introduced into the body. There are, of course, immediate social and ethical questions about the regulation and use of such technologies and how they might feed off the prejudices that already divide and destroy societies, but, even more ominously, it is impossible to predict their long-term effects on human evolution.

Many experts, however, expect the next big change in our evolution to arise from sophisticated mechanical or electronic modifications. Space travel made demands on the human body which were beyond its natural capabilities. Clothing, equipment and the body itself all had to be adapted for survival in the new environment, and this gave rise to the notion of the cyborg, a term first coined by Manfred E. Clynes and Nathan S. Kline in 1960 for a human-machine hybrid that could survive in an extraterrestrial environment. The first syllable of 'cyborg' derives from 'cybernetics' (from the Greek for 'steersman'), which is the study of control systems and comparisons between artificial and biological systems. The second syllable comes from 'organism', which adds emphasis to the significant part that will still be played by the human being who acts as host to the technology. The cyborg's abilities extend beyond human limitations by mechanical, electronic or chemical means incorporated within the body. Like evolution, cyborg technology provides the elements necessary for genetic isolation and speciation. Although like recognizes like in order to maintain the genetic identity of each species, Darwin acknowledged that there are exceptions – there is a small percentage of hybridization, mutation and other forms of genetic drift.

A strong motivation for the development of the cyborg is the desire for eternal life, a desire which must be older than history, linked as it is to the instinct for survival. Myths ancient and modern tell of human beings who want not only to hold dominion over all the earth but also to be god-like in their immortality. But there would be terrible disadvantages in living for ever, and few have considered them carefully. Moreover, cycles of birth and death are the basis of the natural order, and we have a responsibility to make room for succeeding generations.

Besides our instinct for survival, human beings also have a constant desire for more – more of everything – land, possessions, pleasure, stimulation. 'Greed is good,' said Gordon Gekko in the film *Wall Street* (1987). Early explorers and astronauts were searching for new territory as well as a test of the limits of their physical bodies. We have moved from skis and skates, telescopes and spectacles, hang-gliders and diving apparatus to the medical prostheses, procedures and implants that enhance our bodies' powers and make up for many of their deficiencies. Alongside these developments, new technologies are creating a hyper-reality to heighten our experience of the world around us.

04 + 05●●
Just as engineers look to nature for
inspiration, designers reverse the
process and fantasize how man-made
technology could affect nature.
Here, the Swiss ceramic company
Bopla! have designed a range of
tableware using animals and birds,
including parrots, that appear
to be part-organic, part-machine.

04                                    05

The introduction of cyborg-related products has
forced designers to think very carefully about
how the consumer will use such interactive
products as well as the more basic question
of why they would want to use them. Some
have appeared as gimmick products but quickly
disappeared again, and it is obvious that the
consumer wants more than just the latest
gadget. The new technologies have made
technologists, designers and manufacturers
work together instead of in isolation. What is
now emerging is a more human-centred type
of product design which is easier to use and
relate to – in other words, 'user-friendly'.

The essential difference between this and
earlier stages in human evolution will be that
we are going to be able to choose how we want
to change. But this choice may not remain open
for long, and we may not make it wisely. As we
enter the cyborg age, we must decide which
elements to embrace and which to reject.
Many of the choices will be based on ethical
questions affecting both the rights of the
individual and collective humanity. We also
have to consider the threat that Artificial
Intelligence and Artificial Life would pose to
our very existence if increasing degrees of self-

In the mid-twentieth century the appearance of the robot of the future was open to many different interpretations. This bronze *Robot* by sculptor Eduardo Paolozzi is dated 1956 and shows a very organic form. Most depictions of the day tended to be more geometric and technological in appearance to indicate the existence of wires and circuit boards beneath the 'skin' of the artificial life form. The theme of misunderstanding and underestimating the nature of robots has been exhaustively explored in science fiction and film. In the absence of an alternative viewpoint, science fiction has had a major influence on the way that we view the robot as a machine that might take over and eliminate us.

autonomy culminated in control over their own evolution. How soon would it be before human beings became outclassed?

The cyborg technologies in production and under development have the potential to bring untold benefits to humankind. They can bring people together who might otherwise be geographically or socially isolated. They can equally create distance and bring into being a class system based on the technology 'haves' and 'have-nots', corresponding with the rich and the poor both within nations and globally. Some research, in the area of medicine especially, might be stopped on moral and ethical grounds. Most people find it very difficult to accept interference with the genetic make-up of human beings, and it is likely to remain unacceptable to the majority for the foreseeable future. Certainly there are some questionable applications for these advances, but any debate would be better focused on the use or misuse of these technologies as opposed to the development of the technologies themselves.

Science fiction is an acknowledged barometer of the public attitude to scientific advances, but its influence extends much further to affect opinion and legislation on technology. If we look beyond science fiction as entertainment, and consider its role in formulating our attitude towards science and how the future will look, it becomes something very different. Pliny's tales and descriptions of hybrid creatures in his *Historia naturalis* (Natural History) created misconceptions for centuries after his death in AD 79, engendering a fear of unknown peoples and cultures. Mary Shelley's *Frankenstein* (1818) set the agenda for science as the harbinger of man's destruction. While modern science fiction is often based on actual technologies, in order to appear as futuristic as possible it is usually written while these are still under development, and the use and possible misuse is largely unknown. It is often the writer's imagination that takes science to excess, but it can also raise very valid questions about the implications. It is our responsibility, both to ourselves and future generations to map out and control our own evolution●●

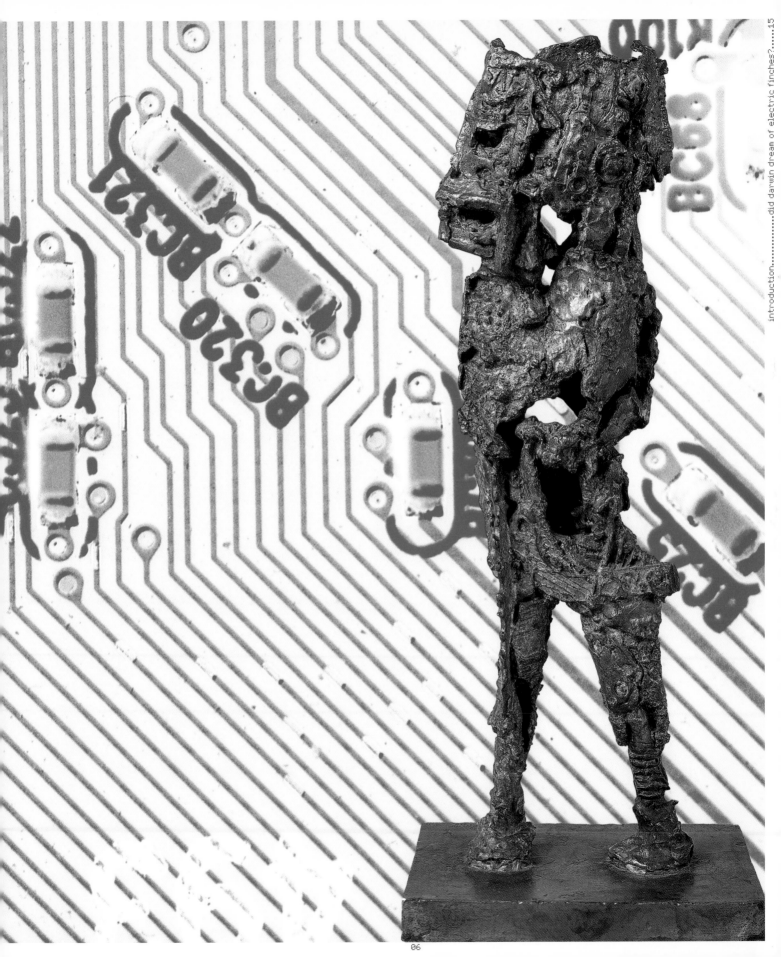

There are a myriad reasons for the pursuit of the cyborg state, from improving the quality of life for the amputee to cheating death by living forever. The amputee wants prosthetic limbs, while death-deniers have their corpses immersed in liquid nitrogen in the hope of one day being brought back to life. The quest for eternal life is an age-old human pursuit and can be traced through every age in religion, alchemy, the sciences and the arts. It is a concept that is deeply embedded in our psyche, and is exerting a powerful influence on the scientific development of the cyborg.

Just as there is no one viewpoint of how to achieve eternal life, there is no single opinion on what constitutes utopia. One person's utopia is another's dystopia. For its residents, Disney's Celebration town in Florida is a peaceful haven where the streets are clean and citizens feel safe. The price (apart from the cost of the real estate) is a range of restrictions on everything from the colour of the front door to the ornaments that are visible to passers-by. In contrast, the bustle of Hong Kong's Nathan Road with its abundance of neon signs, electronic goods, restaurants and people, appears utterly chaotic. To the traders and residents, however, it, too, is a place where they feel at home.

# the human body in eternal life

'Couldst thou make men to live eternally
Or; being dead, raise them to life again,
Then this profession were to be esteem'd.'

Christopher Marlowe, *Doctor Faustus*, 1588

Although the desire for eternal life is shared by many, few who seek it have any clear idea of what to expect from such an existence. Technology can help us to have a longer and better quality of life, yet longer and healthier lives are not enough for some of us: we want to cheat death and live forever. Spiritual and secular thinking are broadly in agreement about the benefits of eternal life. Principal among these is the opportunity of reaching an ideal state of existence, whether mentally, spiritually or even in a physical place. But in the pursuit of eternal life through the cyborg state, it is far from clear whether the outcome would be a utopia or dystopia.

Most spiritual thinkers regard eternal life as possible once the limitations of the physical body have been transcended. In the Christian religion for example, the main obstacle is having a physical body at all. The New Testament reminds believers that 'The spirit is willing but the flesh is weak'. This has some resonance with the 'cyborgian' view that the human body is limiting and the weak link in the man-machine loop.

In most cultures, respect is shown to the lifeless physical envelope of the deceased. It is usually disposed of through burial or cremation and associated rites of passage. In some societies, the body is believed to accompany the deceased through its journey into the afterlife; in others, the body has to be properly disposed of before the soul can move on peacefully to another stage in its existence. Disposing of a corpse may involve a series of rituals culminating in a final 'resting' or burial from which it is taboo to disturb it.

In Christianity, it is only the soul that is considered to exist in eternity, although the person is believed to stand in judgment before God in his or her human form. The ancient Egyptians, however, took the view that the body would continue with the deceased in eternal life. Mindful of this, the various rituals used in the process of mummification were designed to take care of the deceased's physical and spiritual well-being. The most important of the ancient Egyptian rituals is the 'Opening of the Mouth' ceremony originating in a practice believed to endow statues with the ability to support a living spirit. This was adapted for funereal purposes to restore the corpse's ability to see, hear, breathe and sustain nourishment, allowing the deceased to pass into the afterlife retaining bodily functions.

The Egyptians believed that humans were composed of physical and non-physical elements. The physical included the body and the heart, and the spiritual included the *ka*

01

02

03

04
05

03••
In this eighteenth-century
manuscript, the goddess Lakshmi is
shown feeding the food of the
immortals to her consort, Vishnu.
They are both sitting on a fully
blossomed lotus, the seat of divine
truth. Lakshmi is the goddess of
prosperity, purity, chastity and
generosity, and her four hands
represent four spiritual virtues.
Her aura is considered one of divine
happiness, mental and spiritual
satisfaction, and prosperity.

04••
In Greek mythology, Aphrodite,
jealous of Psyche's beauty, sent
Eros to make her fall in love with
an unworthy man. Instead, he himself
fell in love with Psyche, visiting
her nightly until she discovered
his identity. Eros fled and Psyche
suffered the wrath of Aphrodite,
but then the lovers were reunited
through the intercession of Zeus.
In order to marry Eros, Psyche was
made immortal. The lovers are shown
in this painting by Jacqueline
Morreau entitled *Psyche and Eros,
Shirt of Flame*.

05••
Nosferatu, played by Max Schreck
in the 1921 classic vampire film,
is indeed a 'symphony in terror'.
The atmospheric cinematography
emphasizes Schreck's hawk-like
appearance, and his fingers are
grotesque talons. The undead
cannot be considered human.

06 + 07••
The mummification of the deceased
body reflected the Egyptian belief
that preservation of the corpse was
necessary for eternal life. The *ka*,
a double accompanying a person from
birth through death, represented
spirit and intellect, and was
symbolized by upraised arms (this
example is from the Egyptian Museum,
Cairo). The internal organs were
removed, though the heart was later
replaced, the body was dried and
resin applied to exclude moisture
before the wrapping process.

08••
Containers stored the organs of the
mummified body. Each of these (from
the British Museum) bears the head
of a protector: the ape of Hapy, the
dog of Duamutef, the human of Imseti
and the hawk of Qebehsenuef.

(a double that accompanied a person from
birth) and the *ba* (the spirit that left the body
in the form of a human-headed bird), regarded
as essential to the survival of the individual in
the afterlife. During mummification most of
the organs were removed from the body, the
exception being the heart which was retained
for judgment by the gods. Removed organs
were interred separately in Canopic containers.
The preserved corpse acted as host to the *ba*,
which was thought to perish without the
mummified body to return to. This aspect of
Egyptology has served science fiction writers
and filmmakers well throughout the years.

For the vampires of fiction it was their coffins
that were safe refuges for their bodies.
Bram Stoker in his *Dracula*, published in 1897,
decreed that the undead must reside in their
coffins during the day, away from the light of
the sun. *Nosferatu: A Symphony in Terror*,
directed by Friedrich Murnau and released
in 1921, was the earliest screen version of
Stoker's story. Nosferatu, played by Max
Schreck, went to great lengths to ensure that
his coffin accompanied him on his journey to
England. The desecration or destruction of this
resting place resulted in instant decay and
death. Death in this instance came as a release

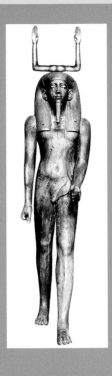

06    07    08

from what had become the 'curse of immortality' to all but Dracula himself.

Vampires aside, it is customary to bury a body intact, although there are exceptions to this rule. The body parts, the hair and the possessions of those considered to have lived exceptionally good lives – the saints and the prophets – can be found scattered around the world, venerated as holy relics. Marina Warner in her book *Alone of all her sex – the myth and cult of the Virgin Mary* (1983), notes that, were all the breast milk attributed to the Virgin in church relics genuine, by her calculation the Virgin would have produced the same volume of milk as an exceptionally high-yielding cow.

In the West, relics of Christian saints are kept in elaborate reliquaries, often in the shape of a casket or the holy body part itself. In the medieval room at the British Museum in London, a thirteenth-century German relic of St Eustace is displayed. The reliquary is in the form of a head made from silver gilt and encrusted with stones. Underneath, a wooden core has been hollowed out to contain a parcel of relics that include part of the saint's skull. It was believed that the proximity of the bones or the belongings of the saints could impart

their essence to believers , and increase their hopes of eternal life.

Among several west and central African peoples, human hair is often added to a mask or sculpture. This has a particular power derived from its source – a venerated ancestor or a brave warrior. Sometimes the hair is styled to add further significance – symbolizing ancient wisdom, for instance. Such objects are used in rituals and healing ceremonies when their power is evoked by diviners or healers.

The twentieth-century artist, Alberto Giacometti, spent his life trying to communicate the essential nature of the individual through painting and sculpture. His sculpture shows skeletal figures that seem to have almost fully decomposed yet are recognizable despite the absence of detail. For the artist the physical body became individual identity, as unique as each person's fingerprints. The cosmetic industry has also tried to capture the essence of the individual, and produces a proliferation of perfumes endorsed and named after celebrities. The French couturier Marcel Rochas designed stunning dresses with trademark Chantilly

lace for Mae West, then paid homage to her curvaceous figure by basing the design of his Femme perfume bottle on the actress. Paloma Picasso successfully markets her perfume on the basis of her enduring style. When Marilyn Monroe repeated the advertising slogan of Chanel No 5 in the 1950s, 'The only thing I wear in bed is a little Chanel No 5', sales for the perfume rocketed and have remained one of the best-selling scents since. The most successful fragrances are bought by customers wanting some aspect of the style or fame of the celebrity who is endorsing the scent. Perfume is a sort of modern equivalent of the saint's relic, but for only a few dollars it can become a personal possession.

Some people do not think it necessary for the body to be buried intact, and will give consent in advance for their organs to be donated to medical science, often for transplant operations. Corneas, kidneys, hearts, heart-and-lungs are among the body parts that can now transplanted from suitable donors. The science fiction film *Coma* (1978), starring Michael Douglas, is based on a hypothetical situation where the medical profession abuses such a system. The story

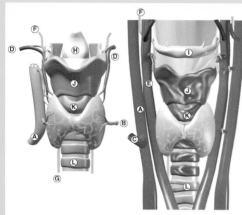
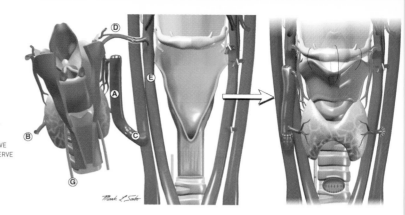

A JUGULAR VEIN
B MIDDLE THYROID VEIN
C COMMON FACIAL VEIN
D SUPERIOR THYROID ARTERY
E COMMON CAROTID ARTERY
F SUPERIOR LARYNGEAL NERVE
G RECURRENT LARYNGEAL NERVE
H EPIGLOTTIS
I HYOID BONE
J THYROID CARTILAGE
K CRICOID CARTILAGE
L TRACHEA

09

**09••**
The world's first successful larynx transplant was performed at the Cleveland Clinic in 1998. Speechless for twenty tears, the patient was able to sing in his church choir just three years after the operation. The team led by Dr Marshall Strome transplanted a donor larynx as well as part of the trachea and pharynx, which allowed the patient to speak and swallow normally.

**10••**
The production of reliquaries, or shrines that contain relics, is an almost universal phenomenon. Many were designed to be portable, such as the Tibetan reliquary, or *ga'u*, commonly made as a miniature church or sarcophagus. In Tantrism and Tibetan Lamaism the bones of holy persons are used to make ritual musical instruments, such as flutes or horns. Reliquaries also take the form of human figures, hands and feet. Shown here is the gold-plated wooden image of Ste Foy at Conques in France, made about 985, and covered with votive gifts from the faithful. In a cavity in the back of the figure, wrapped in silver, is a little skull, said to be of the twelve-year-old girl martyred in 303, whose holy relics were brought to Conques in 866, and were credited with many subsequent miracles.

**11••**
The Makonde of northern Mozambique carve Lipico or helmet masks (shown here) for a boy's initiation as an adult. These usually have realistic human features, with wax scarification and human hair. A different more stylized mask is produced in central and western Africa. The Ekoi tribe of Nigeria, for instance, may cover it with animal hide, then further embellish it with human hair, used in order to borrow its power.

takes place in a hospital where simple operations start to go wrong. It eventually transpires that patients are being put into a coma by giving them too much oxygen, then they are stored on a life-support system until their organs can be sold. The sale of human organs is in fact more scandalous than science fiction could begin to predict. In some countries, there are reports of poverty driving people to selling their own body parts, usually one of their kidneys. A young doctor who fled from China reported disturbing details of the trade in human parts there (*Daily Telegraph*, London, 4 July 2001), denied by official Chinese spokesmen. Speaking to Congressmen in Washington, Wang Guoqi said that prisoners were killed to order so that doctors could take body parts for sale to patients from the West (largely to Chinese-Americans) and the Far East. He said that most executions were carried out with a bullet to the back of the head, leaving vital organs undamaged, but when a cornea was needed the victim was shot through the heart. It is estimated that in 2000 about four hundred people travelled to China from Taiwan and Hong Kong alone to receive transplants (the standard rate for a heart or liver is around £30,000 and corneas £8,000). Many were aware of the source of their 'donated' organ.

Xenografts, or animal-heart transfers are considered by some to be a viable alternative to the use of human organs, which are in very short supply if obtained legally. PPL Therapeutics and other companies are looking at the possibility of knocking out a gene in genetically modified pigs so that their hearts and livers will be suitable for transplant into humans, and in January 2002 it was announced that five such piglets had been born on 25 December 2001. The heart of the pig is close in size, weight and mass to that of a human heart, though there are concerns about patient acceptance. Drugs can assist in making the pig's heart compatible with the human host, but some patients reject xenotransplantation on moral and ethical grounds. There is also a risk of infection by the porcine endogenous retrovirus which is incorporated into the pig DNA and passed from one DNA to the next. The virologist Professor Robin Weiss has sounded a warning: 'Xenotransplants do not seem to pose a big risk. But then BSE or HIV were not thought to pose big risks when they were first discovered. We obviously have to be very careful.'

A new transplant procedure is being pioneered by the Cleveland Clinic in America which is designed to help patients whose larynxes (voice-boxes) have stopped working as a result of cancer or injury. The operation involves the transplant of a donor voice-box and stimulating the repair of nerves in the patient to enable the vocal cords to work properly. This type of transplant will be unique in that it will also transfer identifying characteristics, something that does not even happen with corneal transplants. The tone and pitch of the voice will remain the same, but the brain governs vocabulary and accent. For this reason surgeons will check in advance that the recipient does not originate from the same area as the deceased, otherwise the recipient would have a voice and accent similar to the donor – most distressing for relatives of the deceased if their paths should cross.

10

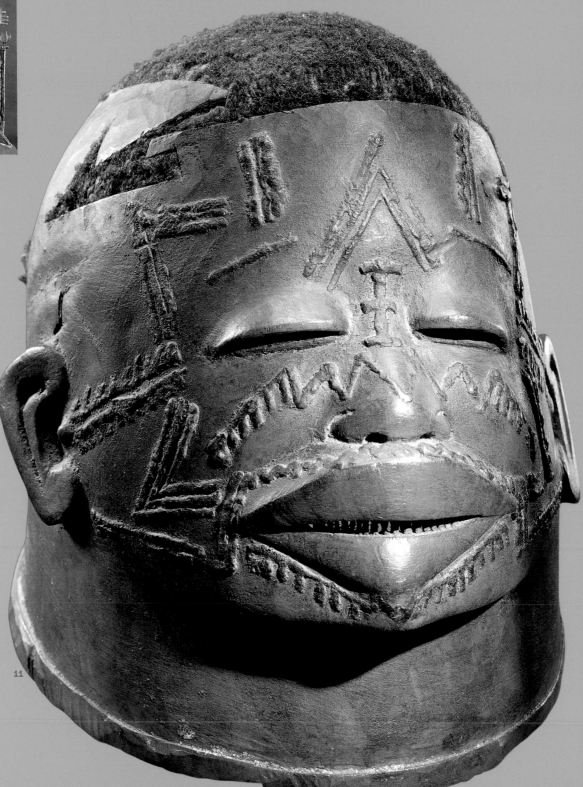

11

**12●●**
Buddhism revolves around a belief in
a continuous cycle of death and rebirth
and is based on the teachings of Buddha,
or 'Enlightened One'. This bronze
Buddha weighs an impressive 200 tons
and is still an important place of
pilgrimage for Hong Kong Buddhists
today as well as the focal point of the
Po Lin Monastery on Lantau Island.

**13●●**
The Egyptian Pharaoh Ramses II was
both an alchemist and a pragmatist.
He did not rely on the alchemist's
art to gain immortality, however, but,
as one of the most prolific builders
of the time, built many works as
monuments to himself, as seen in this
example at the British Museum, thirteenth
century BC. This Pharaoh was further
immortalized in Percy Bysshe Shelley's
poem *Ozymandias* (1818), which mocked
the vanity of self-glorification:
he describes 'two vast and trunkless
legs of stone' in the desert sands,
and 'a shattered visage',

'And on the pedestal these words appear
"My name is Ozymandias, King of Kings:
Look on my works, ye mighty, and despair!"
Nothing beside remains.'

**14●●**
Cornelius Agrippa was a European
alchemist in the Middle Ages who
fell from favour when he disclosed
secrets of alchemical practice. He
is satirized in Dr Heinrich Hoffman's
*Struwwelpeter* (Shock-Headed Peter)
of 1844, cautionary tales for children,
and here is shown dropping naughty
children into an inkwell so that they
emerge a uniform inky black.

'Then great Agrippa foams with rage,
Look at him on this very page!
He seized Arthur, seizes Ned,
Takes William by his little head;
And they may scream and kick and call.
Into the ink he dips them all;
Into the inkstand, one, two, three,
Till they are black, as black can be.'

12

## reincarnation and reanimation

An endless round of rebirth, or *samsara*,
has been a widely held belief in India since the
time of the Upanishads. It is combined with
the idea that it is the duty of each individual
to try to break this cycle and attain a state
of transcendant freedom. Reincarnation
(rebirth in a different body) is a central tenet
of Hinduism, one of the world's oldest religions,
which looks to several reincarnations on
Earth before reaching an ideal state that is
determined by how the person lived in previous
lives. The continuous cycles of death and
rebirth are aligned to the cyclical aspect
of nature, of which humans are very much
a part. In Hinduism, there are four castes:
priests, ruler-warriors, merchants and
farmers, and servants. The castes determines
how far individuals may better themselves,
with the outcastes excluded altogether.

Buddhism also revolves around a belief in
continuous death and rebirth, though crucially
without the discrimination of the caste system,
and is based on the teachings of the Buddha,
or 'Enlightened One' who was born in the sixth

13

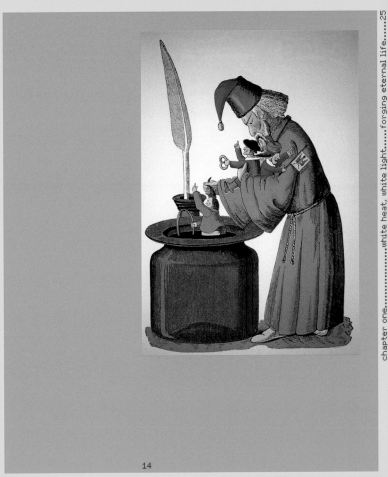

14

century BC. He taught that a release from suffering and the attainment of bliss could be reached by non-attachment to the physical world and an accumulation of positive karma. This was the state said to have been achieved by Buddha at his death.

Reincarnation describes the transmigration of souls from body to body, but the practice of alchemy, the precursor to modern science, included the quest for eternal life, preserving body and spirit together. At one stage in alchemy's development, the Chinese emperors were persuaded that mercury, taken internally, was the secret to living forever. Naturally there were deaths from mercury-poisoning.

Modern science has also been known to kill instead of cure, and many of its early practices have appeared to be both mad and sinister. Mary Shelley's novel *Frankenstein* was written in 1818, and reflected the deep fears of a society in which many scientific experiments relating to the human body were carried out in secrecy.

The nineteenth-century medical profession largely relied on the provision of the corpses of condemned men, though an illicit trade in exhumed cadavers was rampant where demand exceeded supply. Shelley writes of a Dr Frankenstein and his evil creation, a monster pieced together from human body parts and reanimated through the harnessed energy of lightning. The possibility of bringing the dead back to life was being explored through experiments with galvanism (see Chapter 3) at the time that Shelley wrote the book, and she would have been aware of its development. Although committing several murders during the course of the book, the reanimated being is itself depicted as a victim. The true monster is shown to be its creator and it is the name 'Frankenstein' that sits in popular consciousness as synonymous with horror. The cinema has contributed to this merging of the two, spawning a host of horror films in the 1930s inspired by the book, with titles including *Frankenstein*, *Bride of Frankenstein* and *Son of Frankenstein*. The human Frankenstein, a man of science without morality or ethics, has become the source of fear.

America's H. P. Lovecraft is one of many authors to expand on Mary Shelley's reanimated monster. His short story titled 'Herbert West: reanimator' was published in 1922, and more recently inspired *Reanimator*, the 1984 cult classic horror film. In the story the soul is dismissed as a myth, leaving the only serious obstacle to creating life the procurement of newly deceased bodies that are far from happy to find themselves brought back to life. While Shelley's monster had been assembled from a patchwork of available body parts, Lovecraft's was created using just one body which could be reanimated by injecting chemicals into it: 'I can still see Herbert West under the sinister electric light as he injected his reanimating solution into the arm of the headless body. The scene I cannot describe – I should faint if I tried it, for there is madness in a room full of classified charnel things, with blood and lesser human debris almost ankle-deep on the slimy floor, and with hideous reptilian abnormalities sprouting, bubbling, and baking over a winking bluish-green spectre of dim flame in a far corner of black shadows.'

15

16

Such scenes of Gothic horror are enough to deter most from wishing to prolong the lifespan of their physical body beyond its natural cycle, but there are those who are more than willing to subject themselves to unproven technologies in the hope of one day being reanimated.

Cryogenics is a technology which scientists hope will help to gain immortality for the body. Although it has historical parallels in Egyptian embalming, cryogenics really began in 1964 with the publication of *The Prospect of Immortality* by Robert C. W. Ettinger. Ettinger in turn had been inspired by a story that appeared in a 1931 edition of *Amazing Stories*. It is now possible to cryogenically freeze humans by placing them in a low-temperature environment immediately after clinical death with the intention of reanimation at some point in the future. The head alone or the whole body can be cryogenically frozen. Although the technology exists to freeze and store the body, revival from the cryogenic state has yet to be attempted.

'Has the famous story that stands at the beginning of the Bible really been understood? The story of God's hellish fear of science? ... Only from woman did man learn to taste the tree of knowledge. What had happened? The old God was seized with hellish fear. Man himself had turned out to be his greatest mistake; he had created a rival for himself; science makes godlike – it is all over with priests and gods when man becomes scientific. Moral: science is the forbidden as such – it alone is forbidden. Science is the first sin, the seed of all sin, the original sin. This alone is morality. 'Thou shalt not know' – the rest follows.'

Nietzsche, *Der Antichrist*, 1888

15••
In *Frankenstein meets the Wolf Man*, 1943, the Wolf Man (Lon Chaney) seeks out Dr Frankenstein to put him out of his misery. The doctor is dead, but his creature (Bela Lugosi) has been reactivated. Dr Mannering offers to help the Wolf Man, but tricks him, leading to a showdown between the two monsters. Once more it is the scientist trying to play God who is shown as the most evil.

16••
The computer illustration above is of cryogenic pods. The Alcor Life Extension Foundation in California, already preserve bodies long-term in such caskets of liquid nitrogen until a future technology (cryonics) can resurrect them. As soon as the patient is declared legally dead, a team cools the body and administers intravenous medication to protect cells from the effects of lack of blood flow and tissue decay. Alcor believes that if the brain cells and brain structure are preserved, the person is still potentially alive.

In the 1983 science fiction film *The Man with Two Brains*, Steve Martin despairs of finding the ideal woman until he finds the perfect brain. He then tries to find a suitable body as host to the brain. Cryogenic technology is now being used by scientists to store human organs including the brain. Cultured brain cells can be stored in cooled liquid nitrogen for later analysis and experimentation. Scientists believe that research on brain cells may help them understand the workings of the brain, and also how diseases such as Alzheimer's start and progress.

# uploading

While cryogenicists worry about what will happen to the brain as it thaws out from its frozen state, some scientists are investigating the possibility of persuading the human body to accept a brain transplant. Others look beyond replacing like with like, and envisage a future where the human is empowered with a less fragile, inorganic replacement.

The possibility of transplanting the brain has captured the imagination of science fiction writers and filmmakers for decades. The brain is the place of intellectual activity but many regard it as home to the soul or psyche. The process of transplanting the brain is not without its difficulties, as has been comically pointed out in Steve Martin's *The Man with Two Brains* (1983). In true Hollywood style, Martin is a brain surgeon who desires the perfect

17

18••
In 1966 when the film *Fantastic Voyage* was released showing people moving through the human body, it was seen as yet another fabulous future world imagined by science fiction writers and special effects men at the film studio. By the end of the century, however, it was possible to move through the body using a range of medical imaging technologies, such as the endoscope (a fibre-optic tube). The Australian performance artist Stelarc has used the technology to create his infamous *Stomach Sculpture* (1993). The artist lowered a specially adapted endoscope into his stomach (which tried to reject the 'foreign' device), recording all on video to reveal a series of pulsating images, both attracting and repelling the viewer.

wife, whom he finds – as a brain in a jar of formaldehyde. He then sets about finding an appropriate body. The final result is his choice of brain and body, but the brain has neglected to tell him that she is a compulsive eater. In a final twist, the body of his dreams does not remain so for long.

Actual brain-transplant experiments are, of course, not so humorous. Robert J. White, Professor of Neurosurgery at Case Western Reserve University, Cleveland, Ohio, records the details of a 'successful' mammalian transplant in a rhesus monkey. When the monkey regained consciousness, it displayed its usual wakefulness, aggressiveness and alertness. It only lived for eight days, but the professor's reaction was optimistic: 'With the significant improvements in surgical

18

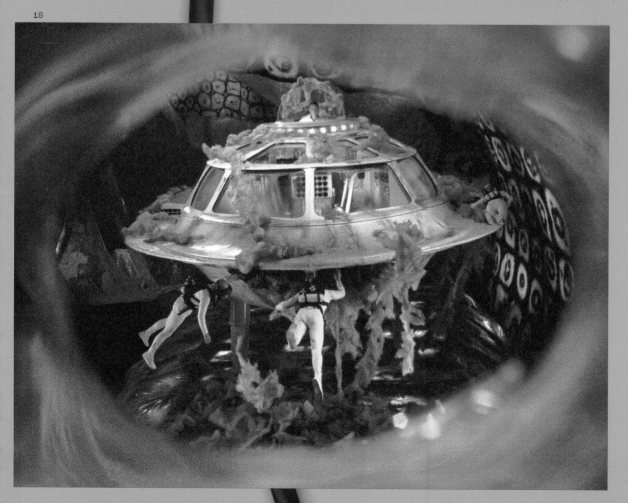

techniques and post-operative management since the [1970s], it is now possible to consider adapting the head transplant technique to humans.'

A pioneering form of brain surgery at Grenoble University Hospital in France offers hope to patients suffering from Parkinson's Disease. The operation was described in an article in July 2001 in the *Guardian* newspaper by journalist Michela Wrong who witnessed the treatment of a friend who was beyond the help of any control drugs. During the eleven-hour operation, electrical activity is picked up by electrodes as they are being inserted into the brain. All is performed without the 'oblivion of anaesthetic' – the brain tissue does not register pain, and the patient must remain conscious throughout the operation to guide the surgeon to the correct location with a running commentary on the sensations that he or she is experiencing.

Replacing the organic brain with a carbon copy machine is known as uploading and is the ultimate challenge for a brain transplant. The benefit of such a system would be to eliminate the weakness of the organic brain which can be injured and is prone to debilitating diseases such as Alzheimer's. The machine brain could outlast its human host and live forever. Several procedures have already been proposed by scientists in America. In one process the subject's brain would be frozen solid, then cut into thin slices and scanned. The computer would then reconstruct the

brain in an artificial substrate that would then be placed back in the subject's body. The first (scanning) part of this process has already been undertaken with a volunteer from Death Row in America after his execution.

Hans Moravec, a robotics and Artificial Intelligence (AI) researcher at Carnegie-Mellon University, Pittsburgh, has outlined the most terrifying procedure for uploading in his book *The Mind Children: The future of robot and human intelligence* (1990). A robot surgeon equipped with billions of nano-scale sensitive fingers would put its manipulator fingers inside the head of the patient, who would remain conscious. The manipulators would then peel away cells and expose layers of the brain, closing blood vessels with clamps. The fingertips would be equipped with electric and chemical sensors to monitor the activity of the exposed brain cells. Once assessed, the robot's computer then configures a machine simulation to reproduce their activity. Cells would then be replaced layer by layer until the organic brain is totally replaced by a machine. Although many nanotechnology manufacturing processes have been proposed, the materials to be used have to be clearly identified.

Fantastic as the description sounds, it is, like many 'futuristic' technologies in science fiction, based on current scientific research and development. In this instance the technology described is nanotechnology which operates at an atomic level, combining the principles

of molecular chemistry and physics with engineering and computer science. The thinking behind the technology was instigated in a 1959 lecture by Richard Feynman at the American Physical Society. The presentation was titled 'There's Plenty of Room at the Bottom'. The premise of his address was that there is no physical reason why atomically precise machines could not be built. In the 1970s K. Eric Drexler took up the idea, eventually setting up the Foresight Institute in California.

One perceived application for nanotechnology is in uploading the human brain. This could be achieved using a series of nanobots to model individual neurons, gradually replacing the organic brain with a machine duplicate. The advantage of this system is that it could be undertaken gradually with no loss of consciousness to the subject. It could also be used on subjects who have been cryogenically frozen, although thawing the brain without literally cracking it would be an additional task to cope with for the nanobots performing the operation. Biology is seen as proof that nanotechnology is a very real possibility. Nature itself utilizes molecular machines that store light energy in the form of sugar, building such immense structures as the giant redwood tree through this process. Initially dismissed by the scientific community, nanotechnology is now being treated as a serious prospect, and has received some funding from the US Government, despite the fact that tangible results are slow to emerge, and at present are largely confined to coatings on existing materials.

# brain-controlled machines

In his 1955 book *Tiger Tiger*, Alfred Bester imagined a world where people would use the power of the mind for travelling. He called this teleportation 'the transport of oneself through space by effort of the mind alone. People could also communicate through telesending which allowed telepathic people to broadcast their thoughts to the world.

Linking the brain directly to the machine is a technology that is already being tried and tested. As early as the 1960s, scientists discovered that certain components of the electrical signals emitted by the brain could be controlled by people themselves. Electroencephalograms (EEGs) can be recorded from the scalp and used to issue simple commands to electronic devices. Until the late 1990s, this technology remained largely a laboratory curiosity explored by the American and British Air Force as a futuristic means for pilots to fly jet planes. There has now been renewed interest from the medical profession which is interested in its possible use to help paralysed and paraplegic patients gain control of their limbs.

Miguel Nicolelis, a neurobiologist at Duke University, North Carolina, is one of the pioneers in researching the use of neural implants to study the brain. The intention is to gain a better understanding of how the mind works in order to build implant systems that would make brain control of computers and other machines possible. Nicolelis refers to

such systems as Hybrid Brain-Machine Interfaces or HBMIs. Together with the Laboratory for Human and Machine Haptics at MIT, he made an important breakthrough in HBMI, sending signals from individual neurons in the brain of a nocturnal owl monkey named Belle to a robot, which used the data to mimic the monkey's arm movements in real time. Ultimately, this type of research could allow humans to control artificial devices designed to restore lost sensory or motor functions using brainwaves. Paralysed patients could control their wheelchair or a prosthetic arm, for instance. Researchers at another US college, Emory University, are also looking at harnessing the brainwaves of paralysed patients. In this instance they use a brain implant with some success to allow patients to move a cursor on a computer screen, proving that the principles of using brainwaves can work.

Kevin Warwick is Professor of Cybernetics at Reading University in Berkshire in England. He uses his own body and that of his wife, Irena, as the basis for his research. In early 2001 he had a silicone chip implanted in his upper left arm and that of his wife. The chips have a power source, tuner and radio receiver, and are surgically connected to nerves in the couple's arms. This gives Warwick the ability to send signals via radio waves and a computer to Irena so that as he moves his own fingers he also moves those of his wife. How the information is received and interpreted will be monitored, there is no guarantee that information will be

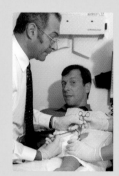

19••
Professor Kevin Warwick is shown having a silicone chip implanted in his arm. His wife was implanted with a similar device linked to his so that they can communicate through the nervous system. In an earlier experiment, using an identifying signal emitted by a chip, he could operate computers, lights and open doors without having to lift a finger.

received exactly as transmitted. There is also the possibility of communicating emotions, since they, too, stimulate nervous activity. The long-term benefit of the research will be for patients paralysed by spinal cord damage where the nerves in the leg below the lesion may still be working but unable to make contact with the brain. This technology could bridge this gap. The effect of the experiment on the couple's relationship also remains to be seen.

The use of EEG signals has already been tested on a human with some (limited) success. Biomedical engineers at Case Western Reserve University in Ohio have succeeded in re-establishing the damaged connection between the brain and body of a quadriplegic patient. Two processes are combined. The first uses a 'neuroprosthetic' Functional Electrical Stimulation (FES), where electrodes are implanted under the skin and used to choreograph movement in the muscles. This is used in conjunction with a Brain-Computer Interface (BCI) which allows the subject to control movement by brain waves alone. Though already of considerable benefit to paraplegic patients, the technology is still at an early stage of development. One of the present difficulties is caused by signal interference where brain cells fire off simultaneously when a message is being transmitted to or from the brain. Researchers are now looking at more invasive brain-computer interfaces that will allow them to tap directly into the motor cortex area of the brain that controls muscle movement.

Researchers at Chicago's Evanston Northwestern University are linking an organic brain to a robot. The living brain of an eel-like fish, the lamprey, is housed in a container of cool, oxygenated salt fluid. The robot is connected to it by electrodes. The fish is drawn by the light-source in its natural habitat and the robot, too, responds to light and moves towards it. Scientists have yet to find a way of keeping the organic brain alive beyond a few days. It is eventually hoped that this technology will help in the development of a brain-controlled prosthetic. As an added twist to the scientist's choice of fish, the lamprey is a vampire.

New manufacturing techniques are behind other developments in this area. Arizona State University has recently developed a class of polymer-based implantable microelectrodes that they are confident will extend the design possibilities for neural implants. Associate Professor Daryl R. Kipke describes the benefit of the BioMEMS technology as including flexibility, adaptable surfaces that can take specially engineered bioreactive coatings, and the use of relatively well developed microfabrication techniques and processes for polymers.

Few people knowingly refuse a life-saving operation or medication, and usually want to prolong their life if they have the choice. Noticeably absent from the promotion of life-extending technologies, however, is any mention of what to expect. For those who have

already taken the plunge, literally, and opted to be cryogenically frozen, it is a blind act of faith. In the event of their being reanimated at some point in the future, they have no idea what they will wake up to. Once given eternal life, what will people do with that vast, endless expanse of time? One scenario depicted by William Gibson in his cyberpunk novel *Neuromancer* (1984) is that we would get bored. He depicts an Artificial Intelligence that, after several hundred years of existence, concludes that it has seen and done enough. It is bored and begs to be allowed to die, to commit suicide. In Greek mythology, Zeus, at the request of Eos, granted Tithonus immortality. Unfortunately, Eos neglected to ask for eternal youth. As Tithonus grew older, he also became increasingly decrepit. Eos soon tired of him and locked him into her bedchamber where he turned into a cicada.

To opt for eternal life would go against the very laws of nature, which provides us with its cycles of birth and death to which our human bodies and our psychology are attuned. And would everlasting life be for all, or only those who could pay for it? Would we tire of ourselves, of our family relationships? Would it mean a world without children, or would there have to be 'culls' of the most aged? But human beings are adaptable, and once given the option of eternal life, might find it difficult to refuse. It may be best to prepare ourselves to enter uncharted territory, as a post-human species••

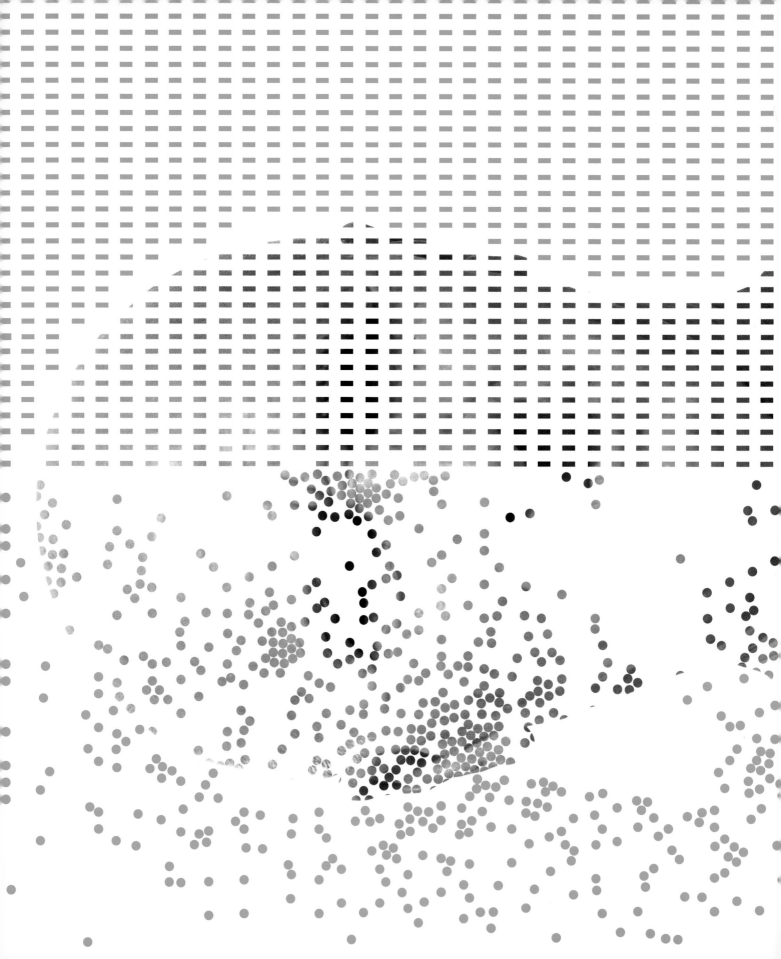

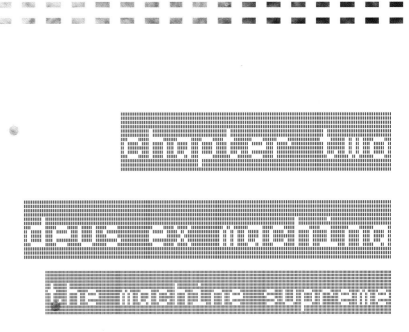

The intended relationship between human and machine
is one of master and servant, with the human as master.
The machine has usually been physically stronger than
the human, and its ability to store and calculate vast
quantities of data has long outstripped the capabilities
of the human mind. The development of thinking machines
and the very real possibility of self-replicating
machines question how much longer it will be possible
to maintain the master-servant status quo.

## 01••

According to the beliefs of the Maya, the gods originally tried to fashion humans using a combination of mud and wood. When this proved unsuccessful, they turned to yellow and white corn (maize). The Maize God is considered the representative of the Maya ideal of beauty, and his long profile and elongated, flattened forehead with partly shaven head and eyebrows also give him a resemblance to a ripe ear of corn.

## 02••

The Book of Genesis tells that God created Adam from the dust of the ground and Eve from one of Adam's ribs. They were given the idyllic Garden of Eden in which to live, with only one restriction, that they did not eat any fruit from the Tree of Knowledge. Tempted by Satan disguised as a serpent, Adam and Eve succumb, and then feel shame at their own nakedness. In this painting by Jacqueline Morreau, an emaciated Adam and Eve are shown coiled in a protective foetal position to form an apple – the fruit which had caused their banishment from the garden and their relegation to the status of mere mortals.

## 03••

William Blake's vision of creation, *The Creation of Adam* (coloured monotype, 1795, Tate Britain, London) owes much to Gnostic imagery which subverts Christian imagery. Instead of causing Adam and Eve's downfall, Satan opens their eyes to show them their real nature. Once aligned with the cosmos, they escape the confines of a separate existence.

01

Although we think of ourselves as human born of human, there are many alternative theories of creation. Religious beliefs and mythologies give various accounts of man's creation from non-human parentage as well as inanimate matter. According to Maya beliefs of the classical period (AD 250–900), the gods originally tried to create humans from mud and wood. When this proved unsuccessful, they turned to yellow and white corn, or maize, with more success. The Book of Genesis tells of God creating Adam from dust, while the Greeks described the Titan Prometheus as combining clay and water to form man. Mary Shelley's story of an artificial life, *Frankenstein*, paid homage to the Greek story with its subtitle *The Modern Prometheus*.

Robert Graves in *The Greek Myths* (1955), reminds us that the Greek philosophers were keen to make a distinction between Promethean and earth-born man, whom they considered inferior. The distinction is also made in Genesis where men were sons of God and women the daughters of men. The inference in both cases is that to be human, born of another human, is by no means an ideal state and we might aspire to better.

02　　　　03

04••
In classical mythology, the three
Fates (Greek Moirai) were the
personification of the inescapable
destiny of man. Klotho was the
spinner who held the distaff,
Lachesis was the apportioner who
drew off the thread, and Atropos –
'inflexible' – the one who cut it
short. Atropos is shown here in a
painting by Jacqueline Morreau.

05••
In sharp contrast to his paintings
of the Spanish nobility, much of
Goya's vision of humankind was as a
cruel and depraved species, and he
sometimes portrayed people as only
half-human. In his *Comida de las
Brujas* (The Witches' Kitchen) the
room is littered with human skulls
and the corpses of babies, and the
perpetrators of this cannibalistic
scene are dog-headed humanoids.

# altered humans

The ability to change into another 'state' appears over and over again in mythology, folklore and popular culture. In Homer's *Odyssey*, the polymorphous sea god Proteus knew all things and could change shape at will to avoid answering questions. Dionysus was said to incorporate the calendar emblems of the tripartite year, having been born in winter as a serpent, turned into a lion in spring and then killed and devoured as a bull in midsummer. Jealous of Poseidon's infatuation with Scylla, his wife Amphitrite threw magic herbs into her rival's bathing pool, transforming her into a barking monster with six heads and twelve feet. These, and other figures from mythology have given way to the more contemporary iconography in the 'super heroes' of Marvel Comics. The Greek myths told of the magical powers of the gods, while modern comic-strip characters, although inspired by Greek and Norse legends, rely on technology for their superpowers. The Marvel Comic character Thor is based on his namesake in Norse legend, while Reed Richards of the Fantastic Four can change his shape, Zeus-like, to any form he chooses. Metamorphosis, whether in mythology or popular culture, is generally taken to be a metaphor for the transformation of identity.

Some transformations do not rely on a physical alteration of the body, but use clothing and masks instead. Both the Copper Inuit and Lapp shamans turn themselves into 'wolves' where the shaman imitates the animal's cries or behaviour. According to the historian and philosopher of comparative religion, Mircea Eliade, the term 'shaman' comes from the Russian Tungusic shaman, with shamanism referring to a 'technique of ecstasy' (*Shamanism*, 1966). Animal forms, such as the wolf, represent helpful spirits to assist in the shaman's journey to the sky and the underworld by allowing him to forsake his

06••

This oil painting shows a Native American Medicine Man, or shaman, performing a healing ritual using a bearskin as part of the ceremony. When the shaman enters into a trance, his soul is believed to leave his body and ascend to the sky or descend to the underworld. By evoking a helping spirit in animal form (shown in the use of masks or skins), the shaman demonstrates his ability to forsake his human condition and his ability to 'die'.

07••

Since Paleolithic times the shaman in every culture seems to have been a combination of magician, healer, priest, mystic and even poet, though some have acted merely as a medium for a priest-figure. A female shaman from Siberia is shown here in a trance induced by her drumming and whirling and perhaps also by the sacred mushroom. She wears animal skins and the wings of birds of prey, and her eyes are covered the better to see through the eyes of the creatures that she is invoking.

human condition. This positive role of the transformation of human to wolf is an exception. (The more common man/wolf depiction is the werewolf, see Chapter 4). The transformation in shamanism is seen variously as his 'double', 'soul in animal form' or 'life soul'. In some instances what happens to the shaman in his altered state can have a bearing on his human form. Within some cultures, when the shaman's alter ego is killed, then he himself dies soon afterwards.

It is one thing to change form but another to render the body invisible, to make matter disappear then reappear. In mythology, the Greek god Hades' most valued possession was a helmet that made the wearer invisible. He loaned the helmet to other gods, including Hermes, who used it to kill Hippolytus. In the collection of the National Museum of the American Indian in New York, is an eighteenth-century shirt from the Pikuni Orsiksika (Blackfeet), highly decorated with glass beads, paint, fur and feathers. Part of the hide has been pierced with a series of half-inch holes which, besides being decorative, have a story to tell. The Blackfeet of Northern Montana tell of a warrior named Big Plume

08••
In Greek mythology the god Hades possessed a helmet that made the wearer invisible. Science fiction often theorizes on the possible effect of this on the human mind. The 1933 film, *The Invisible Man* tells the story of Dr Jack Griffin (played by Claude Rains), a scientist who experiments with invisibility only to find the process irreversible, further transforming him from a mild-mannered doctor to a megalomaniac intent on taking over the world.

09••
David Cronenberg's 1986 film *The Fly* is based on Kafka's 'Metamorphosis'. In this film the young doctor, Seth Brundle (played by Jeff Goldblum), tries to teleport himself by converting his chromosomal programme into computer data. Unfortunately for him, a fly finds its way into the teleportation cabin with dramatic consequences, as shown in this scene where Goldblum emerges from the cabin as a mutant.

who became separated from his war party in enemy territory. Legend has it that a man in a pierced shirt appeared to him in his dreams, and, when the warrior returned home safely, he proceeded to make the shirt that he had seen in his vision. It is said that Big Plume's shirt makes the wearer invisible and protects him from bullets.

In her book, *No Go the Bogeyman* (1999), cultural historian Marina Warner writes of the importance of physical change in relation to identity, 'the seductive invitation of metamorphosis – of turning into something other – has continued to suffuse

fantasies of identity; on the one hand holding out a way of escape from humanity, on the other annihilating the self.' Nowhere has this been better illustrated than in Franz Kafka's story 'Metamorphosis' written in 1915. It concerns a travelling salesman, Gregor Samsa, who one day wakes up to find himself transformed into an insect, a 'monstrous vermin'. His whole relationship with the world around him changes, and he is transformed by what he has become and people's reaction to him. He is rejected by his whole family apart from his sister. He dies and suffers the final ignomiy of being denied a proper burial.

It is left to the cleaning woman to dispose of his remains in the garbage.

There are aspects of cyborg development, such as 'uploading' (Chapter 1) that would confirm Warner's view. Technologists in this area of development consider their work as a means of eliminating the weaker human aspects and replacing them with the machine. But we must not compartmentalize human beings. What may seem a 'weak' link could have a vital function, or provide a necessary component of a combination of parts. Eliminating aspects of the self may in fact lead to a sort of autism caused by technology.

10●●
One of the greatest challenges for today's fighter pilot is combating the effect of G-force, the forces of gravity. A World War I pilot had to contend with just 4gs, the same force felt on a state-of-the-art rollercoaster ride today. World War II pilots were reported to black out at 7gs. These are time-lapse images of Lieut. Col. John Stapp during high G-force testing in the 1950s.

11●●
The modern fighter pilot, with the aid of the latest advances in G-force suits and helmets (a helmet from British Aerospace is shown) can cope with 9gs – nine times the force of gravity. If the G-forces become too much they can overwhelm a pilot and the result is G-lock, a gravity-induced loss of consciousness.

As technology becomes increasingly sophisticated, the role of the human is in many cases in danger of being reduced to that of a facilitator. The relationship between the soldier and his equipment in modern warfare is one instance where this trend can clearly be seen. The modern soldier may be physically located some distance from the combat zone, maintaining a telepresence via computer link. This is an immersive technology where the remote-systems operators feel as though they were physically present through sight, touch and sound. This is achieved through a combination of sensory inputs and feedback, and is used for reasons of safety, cost or physical restrictions. An example is a human teleoperator controlling insect-like robots in such hazardous tasks as bomb disposal or the exploration of planetary surfaces.

Indications are that the future high-technology soldiers will undertake much of their work not on the battlefield but as a remote teleoperator. Even the elite fighter pilots are not immune to the speed of technological development. In the F16 fighter plane a pilot has to withstand a G-force of 9gs, caused by a combination of acceleration, speed and the need to turn the aircraft ever tighter and faster. 'Top gun' fighter pilots, such as the American Thunderbirds, must be physically fit, and have to wear a G-force suit. One pilot described the G-force experience as 'like an elephant standing on your chest.' Blood is forced away from the head and the suit helps to push it back again, preventing G-lock which can cause loss of consciousness.

A new generation of fighter planes is being planned without a cockpit. The role of the pilot will become that of a 'mission manager' potentially located a continent away from the aircraft he or she is controlling. This move is being strongly resisted by pilots who have no wish for desk jobs. The pilotless aircraft will be smaller, more agile and shaped to avoid radar detection. On board will be an Artificial Intelligence in place of the pilot. It is still under debate as to whether the AI or the human will make the decision to attack enemy targets.

Even when the aircraft is piloted by humans, some decision-making is being taken away from them. An AI system is being proposed that would work with the pilot and would monitor health and reflexes by reading brain waves, following eye movements and testing the conductivity of sweaty palms. The computer takes all these factors into account when communicating with the pilot, and may even decide to take over the flying of the plane if it deems this necessary. Pilot and computer would be trained together to enable the two to work well in tandem, much as two human colleagues work in partnership. Chris Hables Gray, in an essay entitled 'The Cyborg Soldier: the US Military and the post-modern warrior', 1989, goes on to describe the possibilities for a direct brain-computer connection. He suggests this might be done through the 'reading' of the pilot's thoughts. In the early 1970s a research project at the Stanford Research Institute, funded by the Department of Defense, goes a step further. Not content with reading pilots' minds, it proposes that the computer should insert ideas and messages into the brain. This raises a number of ethical question both about the rights of the soldier and who is responsible for the actions of this fighting machine.

10

12••
Charles Babbage's Difference
Engine (Science Museum, London)
was a nineteenth-century device
for calculating and printing
mathematical tables. The use
of Jacquard punch cards in
silk-weaving, of chains and
subassemblies, and the logical
structure of the modern computer
all come from Babbage. Assessing
the potential of the computer's
processing ability in 1965, George
Moore noted that each new chip
had roughly twice as much capacity
as its predecessor. This trend
provides a means of calculating
the future exponential increase
in computing power, and is known
as Moore's Law.

# artificial intelligence

During the early part of its development the computer was largely seen as a means of storing and calculating vast quantities of information. There were visionaries who saw the potential in developing a more human-like computer. Alan Turing was one scientist who looked beyond mathematics to the possibility of the computer performing more thoughtful tasks. This approached evolved into what we now call Artificial Intelligence.

The term Artificial Intelligence was coined in 1956 by John McCarthy as a marketing ploy. It was McCarthy who also invented the AI language LISP. By the 1960s machines could solve complicated mathematical problems much faster and more accurately than humans. At the same time, the machine had great difficulty with some easy cognitive tasks, such as stacking random building blocks. Although computers have become more sophisticated, this difficulty with cognitive tasks remains largely unchanged. Writing in the *AI Magazine* (1982) Marvin Minsky suggests the reason for such difficulty is that 'much "expert" adult thinking is basically much simpler than what happens in a child's ordinary play'. The computer has several different methods of problem-solving at its disposal.

Neural networks and chaotic feedback are based on human thinking and are organic processes, while calculation and information storage are considered rational and inorganic. Given this choice of systems, the difficulty in developing a sophisticated AI is not a lack of 'brainpower'. It seems much more likely that the cause lies in something central to human thinking that the machine does not yet have at its disposal.

Self-awareness is the key to our notion of consciousness and what it is to be human. While computers have considerable powers of information storage and processing, they do not, as yet have self-awareness. An AI can be told the difference between right and wrong, given countless examples, and it will draw a conclusion on how it should behave from this information. What it cannot do is understand why its conclusion is the correct one. This is no doubt a programming issue, which goes back to the original function of the computer as a calculating machine. What if a different approach were to be adopted, one that facilitates a degree of supposition? Although self-awareness is essential to our humanness, clinical psychologists regard us as being only very slightly self-aware, with most of what we think about ourselves based on guesswork.

**13••**
Star Wars robots R2-D2 and C-3P0
famously rely on the actors within
for their movement and mannerisms.
Kenny Baker and Anthony Daniels
are the men literally inside
the machines, R2-D2 and C-3P0
respectively. Advances in robot
technology and computer-generated
imagery look set to make actors
as robots redundant in the
near future.

**14••**
Robots are ideally suited to
monotonous or dangerous tasks.
The robots shown here are welding
car bodies on the assembly lines of
the Mazda Car Plant in Hiroshima,
Japan. Most cars in the developed
world are now being produced by
robots, which do not get tired,
hungry, ill or bored, but the more
self-aware robots of the future
may demand more interesting jobs.

**15••**
In science fiction, robots are
often motivated by a human-like
lust for power, whereas in reality,
ironically, self-awareness is
proving very difficult to create
artificially. In Michael Crichton's
1973 film *Westworld*, the story
centres around a games-theme
holiday park with robots. When the
game turns sour, gunslinger robot
(Yul Brynner) leads the hunt to
find and exterminate the humans.

What if we were to take this random element of supposition as a framework model for developing AI? One scenario of the self-aware machine is depicted in Stanley Kubrick's film *2001* (1968). Throughout most of the film we see a utopian view of a future where humans live and work in perfect harmony with technology. HAL is an intelligent machine which speaks in a non-mechanical, human-like way. The relationship between machine and man is social, one of equals rather than that of master and servant. This is fine until the problem of survival arises, when it emerges that HAL has been programmed with a survival instinct that places its own continued existence above that of the human. This is contrary to science fiction writer Isaac Asimov's three 'Laws of Robotics', as quoted in *I, Robot*, 1950:

1   A robot may not injure a human being, or, through inaction, allow a human being to come to harm.
2   A robot must obey the orders given it by human beings except where such orders would conflict with the First Law.
3   A robot must protect its own existence as long as such protection does not conflict with the First or Second Law. (From *Handbook of Robotics*, 56th Edition, 2058 AD.)

In *2001* HAL breaks all three of these laws, leading to the following conversation:

**Dave** Open the pod bay door, please, HAL. Open the pod bay door, please, HAL. Hello, HAL. Do you read me? Do you read me, HAL? Hello, HAL. Do you read me?
**HAL** Affirmative, Dave, I read you.
**Dave** Open the pod bay doors, HAL.
**HAL** I'm sorry, Dave, I'm afraid I can't do that.
**Dave** What's the problem?
**HAL** I think you know what the problem is just as well as I do.
**Dave** I don't know what you're talking about.
**HAL** I know that you and Frank were planning to disconnect me, and I'm afraid that's something I cannot allow to happen.

Many of HAL's attributes are already becoming feasible. Speech and visual recognition and systems that 'read' facial expression are commercially available. Yet these elements are only very rudimentary as yet. There is little 'intelligence' about the way they are being used, which is still largely connected to data storage and linear inputs/output. Roger C. Schank in his essay 'I'm sorry, Dave,

I'm afraid I can't do that' in *Hal's Legacy*, 1997, speculates on whether it may ever be possible for intelligent machines to understand fully what is being said to them. His initial reaction is that it is impossible because of a lack of 'experience'. He goes on to conclude that it would be possible if a computer were given the ability to self-reference and to use powers of association more freely.

The key issues that limit the development of AI have remained the same since its inception: can a machine be self-aware, and could it ever evolve a psyche, or soul? These are generally accepted tenets of what makes us human, and, while such differences remain, there will always be a ceiling to the development of Artificial Intelligence and by extension, Artificial Life.

16

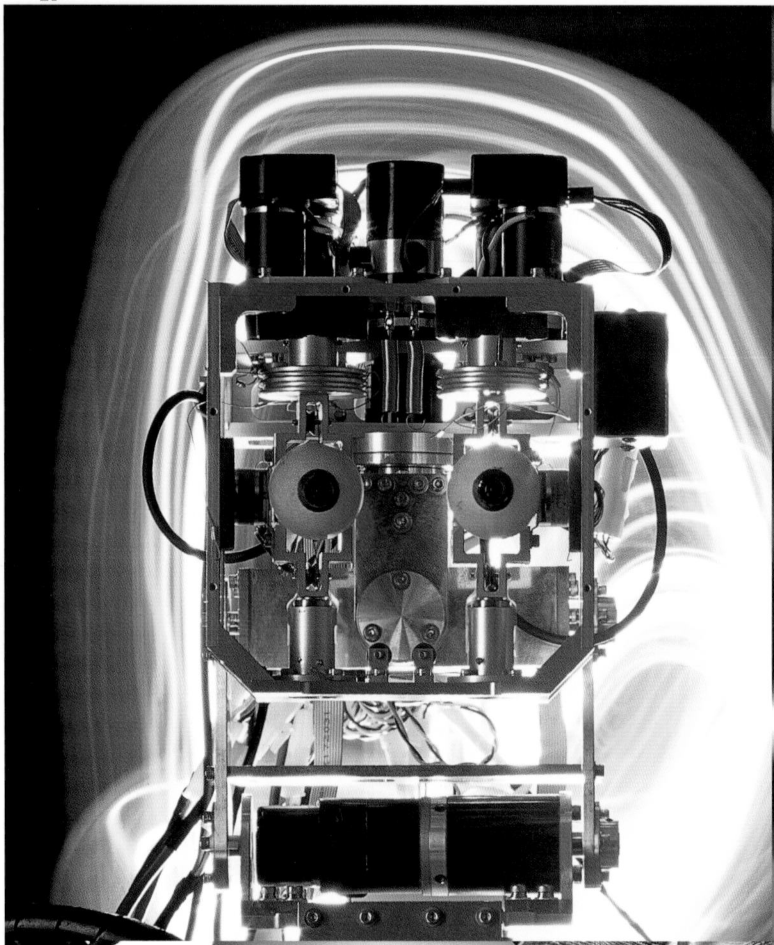

17

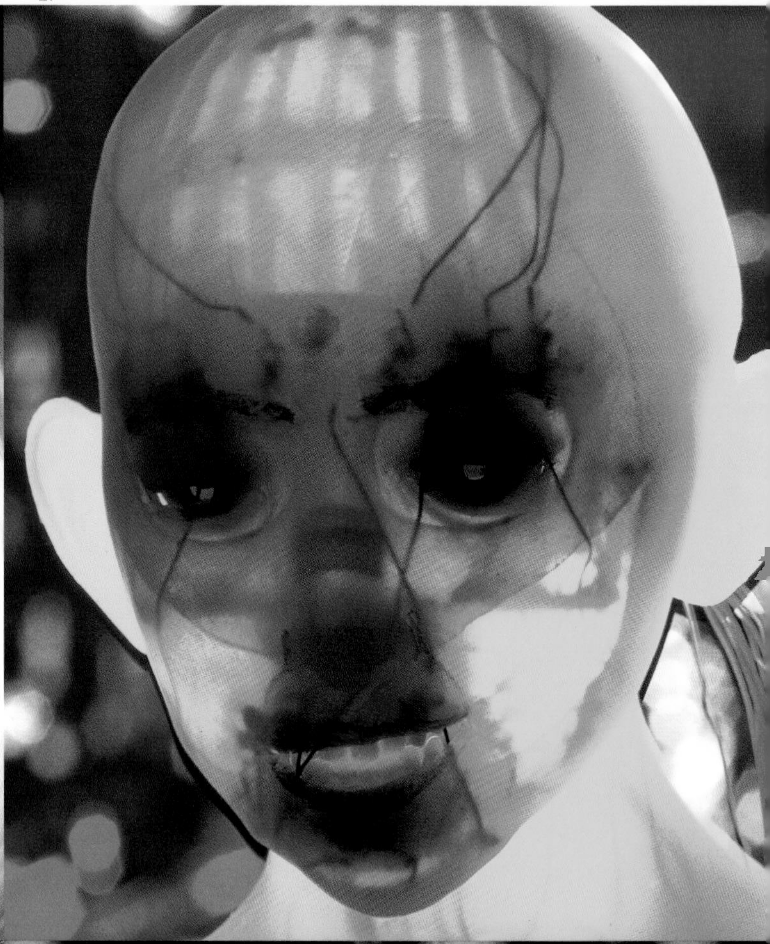

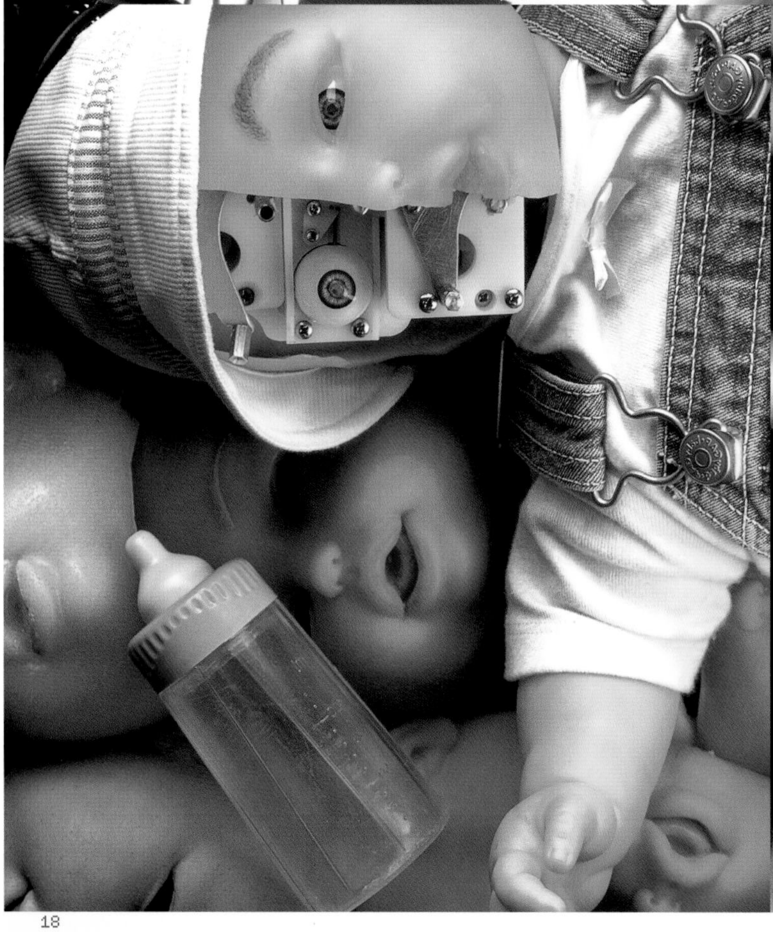

18

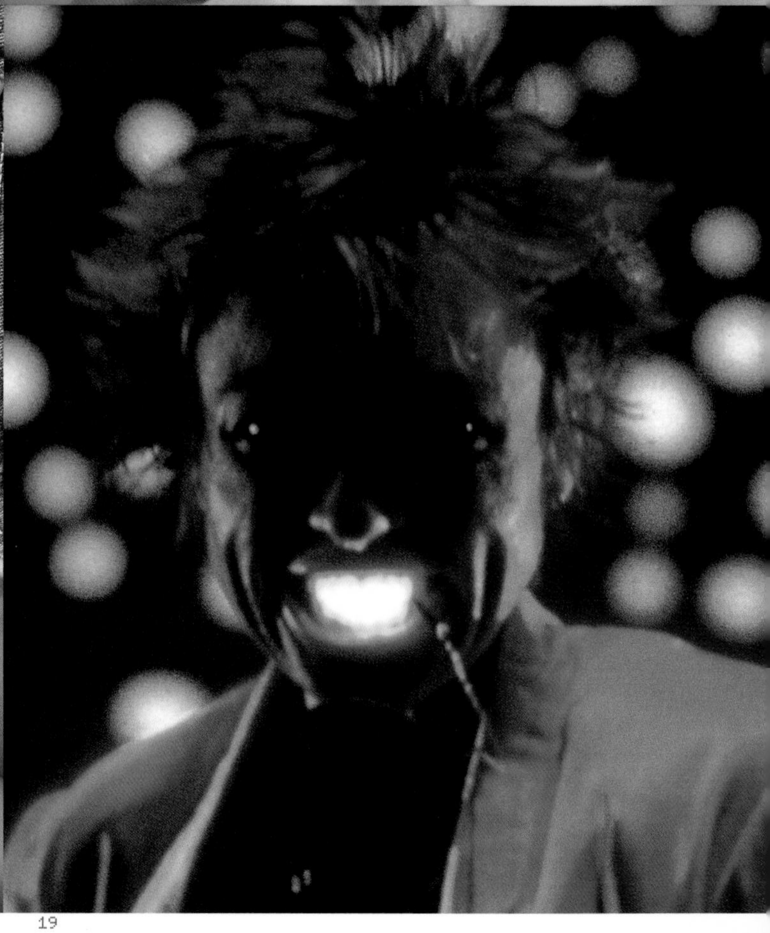

19

20

16••
A number of roboticists are attempting to design robots to move in ways similar to animals, insects, fish and, most importantly, humans. This type of movement, as with Mark Tilden's Spyder or Joseph Ayer's studies of lobsters, offers many benefits over wheels in negotiating difficult terrain. Shown here is WABIAN (Waseda bipedal humanoid), developed by the Humanoid Robotics Laboratory at Waseda University in Tokyo. As its name suggests, it moves like a human, though a little shakily. Some designers believe that robots should seek to copy nature exactly, while others believe it should be only a starting point and needs modification.

17••
Professor Fumio Hara believes that human-friendly robots are necessary 'if they are to coexist with us,' and that they can learn from interacting with humans. Joanna Pransky of Sankyo Robots, the first robotic psychiatrist, is concerned with how robots will respond to humans. Shown here is a second-generation female face robot, developed by Hara and his students at the Hara-Kobayashi Laboratory, Tokyo Science University. The skin is silicone rubber, and a range of systems and smart materials, such as Shape Memory Metals (SMAs) that respond to electricity, allow the robot to change its expression.

18••
Baby IT (BIT) is a prototype of My Real Baby, an interactive robot doll developed by the American toy manufacturer Hasbro and ISRobotics, USA, a specialist robot company set up by Rodney Brooks of MIT. The doll, launched late 2000, can 'learn' simple phrases, and mimic the changing facial expressions of a real baby, giggling when tickled and crying when shaken, but soothed by rocking. When a bottle with a chip inside is placed near the doll, it will suck it. The technology includes sensors, actuators and AI. Cost and the fragility of such systems had previously limited their use as toys for children.

19••
The narrative in performance artist Laurie Anderson's work is often about the body in a machine environment. In *Home of the Brave*, 1985, Anderson placed a microphone and light bulb in her mouth so that she could 'sing like a violin' while her cheeks emitted a blood-red glow.

20••
Central to Ridley Scott's film *Bladerunner*, 1982, is whether a replicant (Sean Young), could ever evolve to be self-aware. In the film the replicant believes it is human, but this also suggests that Deckard (Harrison Ford), who thinks he is human, may not be.

# artificial life

In the seventeenth century, a tea-serving automata, known as a *karakuri* was developed in Japan. Two centuries later Maelzel, a travelling showman, toured a chess-playing automaton known as The Turk around Europe and America. But it was not until 1920 that the term 'robot' was introduced by the Czech writer Karel Capek in his play *R.U.R.* in which robots take over and kill off the human race.

Just as the Bible tells us that God created man in his own image, scientists have looked to create a human-like artificial life. There are two schools of thought on this, both sharing the desire to encourage ease of interaction and understanding between humans and robots. One viewpoint is that the likeness should be physical while another thinks it preferable that the robot should retain an 'otherness' that sets it apart from humans. In Fritz Lang's 1927 film *Metropolis*, the robot Maria was given the destructive elements perceived in the female nature and depicted as a *femme fatale*. This set the scene for subsequent films where machines are depicted as ultimately working against mankind to create a dystopic environment. This is shown most notably in Ridley Scott's *Bladerunner*, 1982, which was based on Philip K. Dick's book *Do Androids Dream of Electric Sheep?* Scott depicts the replicant as human-like to the point of believing it is a human and becoming 'upset' when it discovers it is not. The film has the required Hollywood happy ending with Deckard (played by Harrison Ford) heading into the sunset with the replicant (played by Sean Young). There is a subtext, however, which is the question of whether Deckard is himself a replicant. In the book Deckard considers sparing the replicant's (artificial) life until it casually squashes a bug which is a rare life-form in the book's futuristic world. It is this detachment that convinces Deckard that the replicant is not to be trusted. Its human attributes are only superficial, and it proves that it merely simulates an understanding of the difference between right and wrong.

Professor Rodney Brooks at the Department of Robotics at the Massachusetts Institute of Technology (MIT) is developing a humanoid robot that he has called Cog, derived from 'cognitive'. He is of the opinion that people will prefer to interact with a robot that behaves in a human-like manner. Brooks has already given Cog touch sensors in its torso and joints so that it can interact with people physically without the danger of crushing them. The robot can respond to light and movement and it can also register faces. It can watch a person or object move around a room, but it cannot register what a person is looking at, which is regarded as a key step in understanding what a person may be thinking and is a trait that is also missing in autistic people.

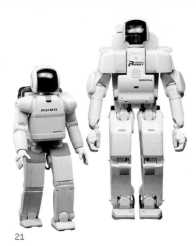

21

22

21••

Honda took the robotics world by
surprise when it announced the
existence of its Honda P3 robot
(right, beside a smaller robot
Asimo). The company view was that
it was simply another form of
mobility which is, after all, its
core business. The project's chief
engineer sees other possibilities
for the robot as a human double
based on the principle of the ninja
who supposedly have the ability to
make an identical self appear - an
interesting proposition given the
current controversy on human
cloning (see Chapter 4).

22••

The performance artist Stelarc
constantly explores the
relationship between human
and machine. In a role reversal,
the human body (his own) is
often positioned as the enabler,
facilitating the machine in his
performance. In *Third Hand*
(1976-81) a robot prosthesis
is activated by electronic
signals from his abdominal and
leg muscles, while his own arm
is moved by remote control
using muscle stimulators.

23••

The ultimate pet, appealing to
adults as much as children, is
Sony's robot dog. The robot's
name, AIBO, is taken from the
Japanese for buddy. Its owners
can activate the robot using a
remote control, and its preset
functions include barking, falling
over and getting up, sleeping
and walking. AIBO has two LED eyes
which will light up red when the
dog is angry and green when it is
happy. Interaction is a large part
of its appeal, with an in-built
video camera and pressure
sensors allowing it to be 'aware'
of and respond to its environment.

While some roboticists are looking at the
possibility of mimicking human thought
processes through AI, others are developing
the technology to allow robots to move in a
humanoid manner. The Honda Motor Company
have developed a biped robot, the Honda P3.
Although it can move and perform tasks in a
human-like fashion, it is pre-programmed
and does not have any AI. However, Misato
Hirose, the Honda P3's chief engineer has
plans to develop a more advanced prototype
that could take on not just human, but ninja
characteristics. Ninjas have the fabled ability
to make an identical self appear, like a
doppelganger. This would enable the busy
executive literally to be several places at once.

Within the scientific community there are
many differing approaches to solving the
problem of movement. In an interview in
the book *Robo Sapiens* (2000), Shigeo Hirose
of the Tokyo Institute of Technology made
the observation: 'If a roboticist sees a man
washing clothes in a river, he would try to
make a humanoid [robot] that could approach
the river and scrub the clothes. If an engineer
sees this, he would make a washing machine,
a very simple rotating machine.' Hirose
himself is something of a pragmatist, sitting
somewhere between the two disciplines.
He works in an area of growing popularity,
biomimetics – the study of good design
inspired by nature. Inspiration is quite
different from mimicry, as his design for
'snake-bot ACM R-1' shows. The modular
design can be bolted together to be as long
or short as needed and moves on nothing more
sophisticated than wheels. The design is being
continuously improved and upgraded with
the hope that a commercially available version
would be capable one day of performing tasks
such as inspecting underground pipes.

The human creator of artificial life-forms is
often reluctant to afford his creation complete
self-autonomy. This desire to play god over
a creation and take on the responsibilities
that come with it are being satisfied by
a computer game *Black & White*, released
in 2001. The objective of the game is to become
the one true god accepted by all, but it reveals
much about the personality of the player, who
has to create an artificially intelligent creature
and train it to be good or evil, to snack on the
villagers or avoid crushing them with each
footstep. The game's author, Peter Molyneux
describes how 'after two hours of play, people
begin to revert to type...after ten hours, what
emerges is a creature and a world that
reflect your personality.'

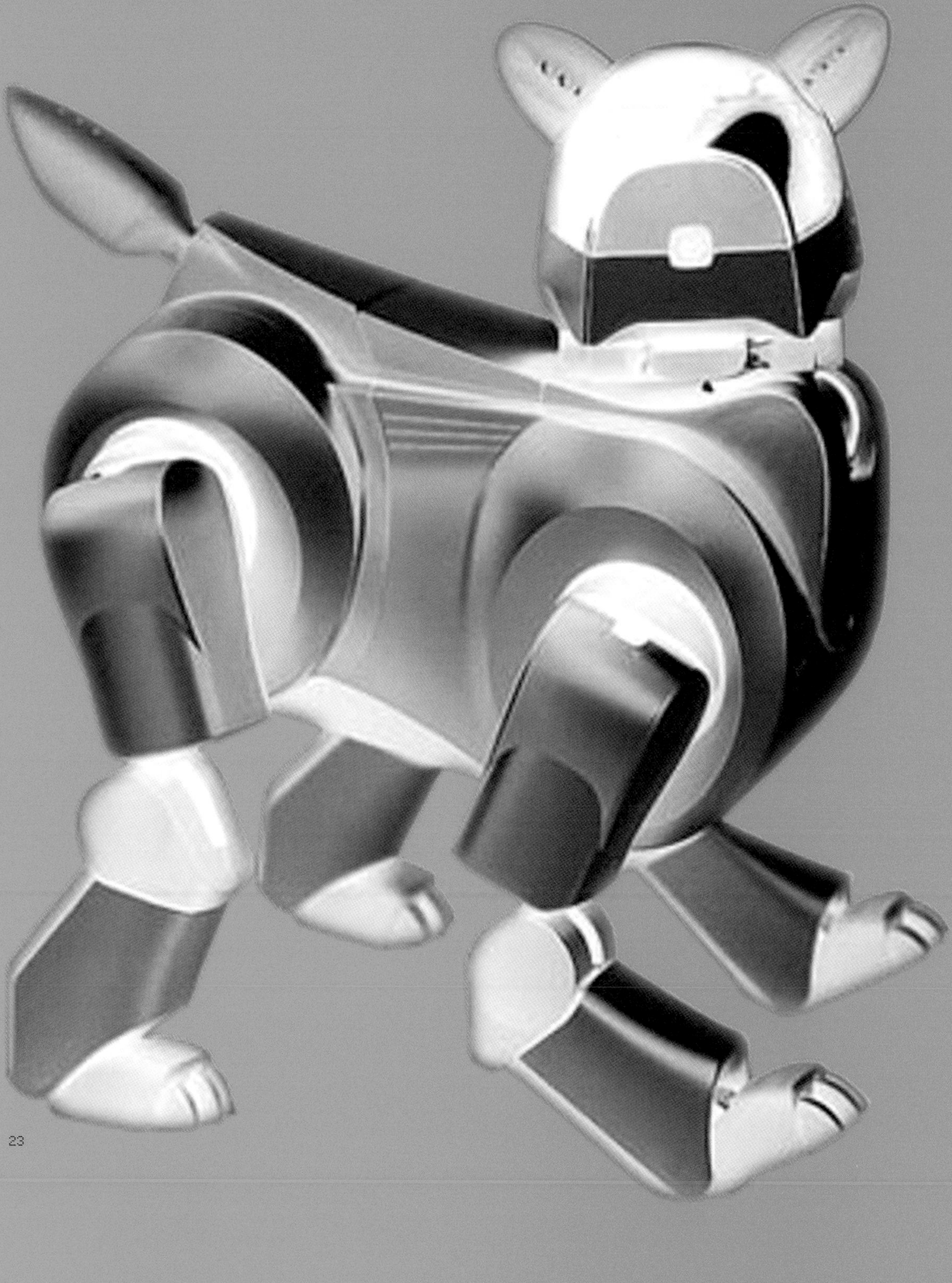

23

# assisting nature

Certain tribal cultures such as the Ibibio
people of Nigeria, held the belief that the
human body in its natural state was
incomplete. Modification was necessary in
order to achieve perfection, hence the ritual
of scarification. In Western Europe, ladies
looked to highlight their feminine attributes.
Clothing emphasized the breasts, hips,
waists and buttocks using steel wires
or bone to structure the garments,
creating extraordinary shapes that were
anatomically impossible. Cosmetic surgery
now allows women to combine a tiny waist
and slim hips with large breasts. This is
not without cost, and Jane Fonda's fitness
mantra of 'no pain, no gain' has taken
on a whole new meaning.

24●●
Scarification and tattoos
demonstrate a pervasive idea
amongst tribal cultures that
the human body in its natural
state is incomplete and requires
modification to achieve perfection.
The painting illustrated here shows
a Maori woman of New Zealand with
facial tattoos, which function as
the identity of the Maori, like
a fingerprint or signature.
In recent years, the tattoo has
also come to be used as a signature
of identity in Western culture.

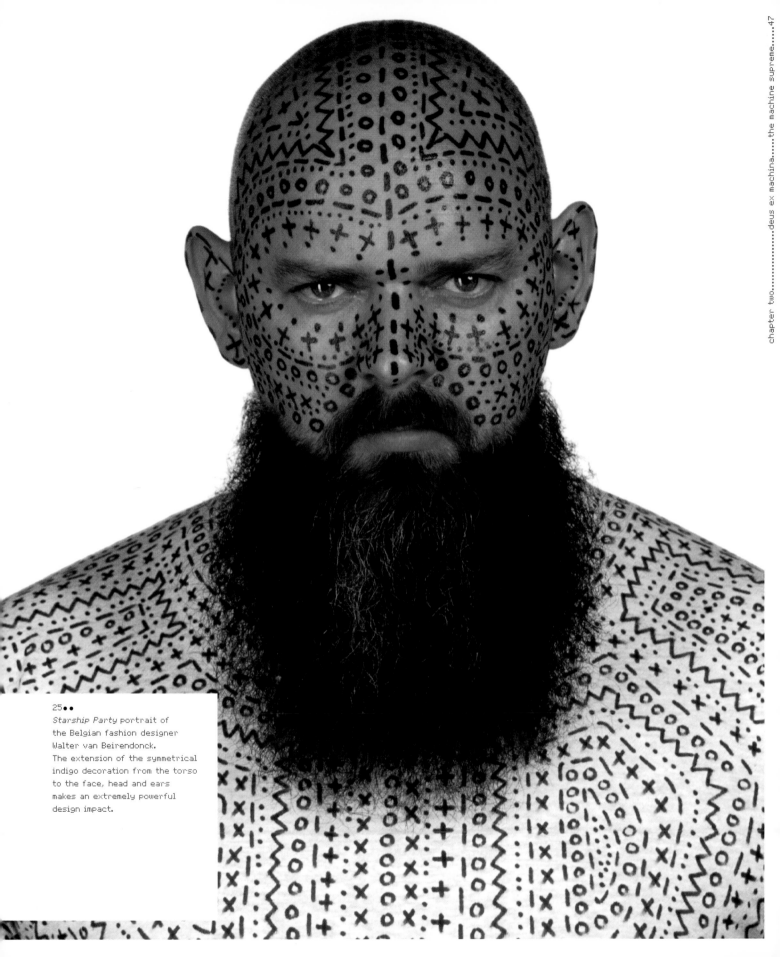

25••
*Starship Party* portrait of
the Belgian fashion designer
Walter van Beirendonck.
The extension of the symmetrical
indigo decoration from the torso
to the face, head and ears
makes an extremely powerful
design impact.

Although cosmetic surgery is a relatively simple procedure, it has not gained widespread popularity. Noses can be bobbed, cheekbones redefined, bagging and sagging nipped and tucked, lips and breasts increased in size, layers of fat sucked out, but it can be painful and expensive, and some procedures don't last for ever. Health concerns have been highlighted by legal action taken over leakages in silicone breast implants. Another factor is the unreality of perfection. Surgically enhanced Pamela Anderson Lee, formerly star of TV programme *Baywatch*, is not considered an ideal beauty by women, nor by many men. She is cartoon-like in her perfection and has become an effigy of womanhood.

The work of French artist Orlan centres on the use of cosmetic surgery and satirizes its role in creating 'idealized' beauty. In *Coil-Hybridizations*, the artist's own surgical alterations were inspired by the false noses and other pre-Colombian and Maya embellishments to the body. Looking to the future and emerging technologies, such as genetic engineering, Orlan is subverting the use of implants by placing cheekbone enhancements on her temples. The result is an other-world appearance that has inspired the Belgian fashion designer Walter van Beirendonck. Orlan became his muse for his 'Believe' collection (Autumn/Winter 1998). Not content with present images of beauty, the fashion designer considers what may

become the perfect human being in the future. The result is distinctly alien, with models given tiny flesh-coloured horns on their foreheads as he explores the world of pre-surgical and prosthetic make-up.

In the words of the satirist Stanislaw Lem in his vision of 2039: 'If prostheticism is voted in… in a couple of years everyone will consider the possession of a soft, hairy, sweating body to be shameful and indecent. …In a prostheticized society you can snap on the loveliest creations of modern engineering. What woman doesn't want to have silver iodide instead of eyes, telescopic breasts, angel's wings, iridescent legs, and feet that sing with every step?' (From *The Futurological Congress*,1971, first published in English in 1974.)

The Greek gods with their enhanced powers were constantly at odds with one another, outwitted not necessarily by a god (or mortal) any greater than they, but with a different cunning. The cyborg soldier is dependent on his technology being better and faster than that of his opponent. Like the computer-games' aficionado, he seeks constant upgrades and power enhancements in an effort to stay one step ahead of the opposition. Whether the perfect body is attainable, or even desirable, is open to debate. It is vitally important for our own humanity that we retain the right to choose how, or even if, we embrace cyborg technologies●●

26●●
In his Winter 1998/9 collection 'Believe' , W. & L. T. by Walter van Beirendonck, the designer explored the world of plastic surgery and changing ideals of beauty. The French performance artist Orlan inspired the collection in her willingness to have her face and body surgically altered as a commentary on ideas of beauty.

Clothing is used both to reinforce and break conventions. It can often perform a symbolic role, emphasizing aspects of the body considered important at a particular time. Illustrated is a heavily pregnant model wearing a Comme des Garçons dress. The image could be compared with Jan van Eyck's fifteenth-century painting known as *The Arnolfini Marriage* (National Gallery, London). It is thought to celebrate the wedding of Giovanni Arnolfini and Giovanna Cenami in 1434, and is usually regarded as a 'marriage certificate' for the couple, though opinions are now being revised. Van Eyck was following a convention in late Gothic art (and also fashion) of exaggerating the child-bearing area of the female anatomy with the drapery of the dress.

27

28

Our senses can now be augmented or enhanced to extend many of our natural human capabilities. And, by implication, the use of such sensory enhancements affects the relationship between people and their environment. The senses of movement, vision, sound, touch, smell and taste can be altered by chemicals, new forms of communication, and hyper-reality devices. Sensory distortions can be synthetic, generating many new experiences that overlap or multiply to create a heightened, or synaesthetic, experience.

At the same time, new technologies are allowing human beings access to new inhospitable places on earth and in space. This has been made possible by the development of advanced portable and protective environments. In tandem, the lifting of human physical and sensory restrictions offers an 'augmented' reality ripe for a new age of pioneering exploration.

The early detection of the presence of others can be vitally important for the survival of a species. In humans sight and hearing are the senses that provide the initial information, while some fish use a highly advanced sense of touch to warn of imminent danger. This Scanning Electron Micrograph (SEM) shows a surface of a neuromast organ in a seventeen-day old zebrafish. The hair bundles at the top of each sensory 'hair' cell emerge through a pore in the skin. Any movement of the surrounding water will deflect the hair bundle, stimulating the hair cell. Fish, as well as aquatic amphibian larvae use such information to detect the presence of other fish or to avoid predators. The inner ears of all vertebrates, including humans, possess similar hair cells that detect sound vibrations.

When the television set first began to appear in people's living rooms, many were concerned that it would be the end of reading, and that children would be permanently damaged by exposure to it. The TV quickly replaced the fireplace as the focal point of the family room, bringing with it a new culture. Two films that explore the power that it exerts over us are David Cronenberg's *Videodrome* (opposite below), 1983, and *The Truman Show* (above), made in 1998. *Videodrome* centres on a gameshow of the same name which consumes and eventually kills participants. In this memorable still, actor James Woods is being sucked into the TV set by the lips of game-show player Debbie Harry. *The Truman Show* looks at reality TV, where everyone except the main character (Jim Carey), are actors in a TV show. Carey is blissfully unaware that his whole life is being televised. The most disturbing characters are the viewers who are *Truman Show* addicts. Once the hero finds out the truth and the programme finishes, the viewers are happy to move on in search of even greater voyeuristic experiences through television.

01

We have always relied on our senses to interact with one another and the world around us. The senses are effectively an interface: through the senses, we absorb and disseminate information. But our natural senses are no longer enough. It is becoming increasingly common to use technology to assist in the way in which we interact with fellow humans and the environment. In cyborg parlance, many of us now live, at least partly, in what could be described as an 'augmented' reality.

Rather like a funfair's hall of mirrors, the augmentation of reality can enhance, distort, or reduce our sensory and physical condition. For example, in medicine the use of prosthetic limbs and implants allows the patient greater personal freedom of movement. New research developments are seeing more direct links being formed between these technologies and the human host. The links allow an increase in sensation via our muscles and nerves that connect the various devices to the brain.

Technology also gives greater autonomy to humans living in extreme environments. Clothing as a technology protects the wearer by functioning as a personal environment. In this context, clothing in the form of a spacesuit provides everything from a micro-climate to oxygen and a communications facility. Such advanced technology acts as a superset of our basic clothing needs, and allows movement and a degree of independence in one of the most hostile places known to humankind: the freezing cold vacuum of space.

Recently, advanced audio and vision technologies have created an explosion in methods of communication, sometimes with bizarre results. Researchers at the University of Lausanne in Switzerland are using Global Positioning Satellites (GPS) to track and monitor overweight patients. A 'logger' device about the size of a mobile phone is carried by the patients, allowing them to be monitored for energy output. This allows a comparison of how many calories are used walking upstairs rather than using the elevator. More seriously, modern communication and surveillance has also opened a running debate about personal empowerment and privacy.

Sensory and digital technologies are creating a whole new environment that has the potential to override normal sensory experience and replace it digitally. Studies show that our senses actually alter in response to a changing environment and also, more controversially, it is suggested that technology might even contribute to evolutionary change. As the technologies become more integrated, we are perhaps approaching a stage where it will no longer be possible to differentiate between the natural and synthesized. This is a state that has already been achieved by Hollywood movies through the use of digital special effects.

The progressive march of technology has the potential to afford greater autonomy to the individual. This was a key theme in Robert A. Heinlein's 1970 science fiction book *Have spacesuit will travel*. In the book, a young man suddenly finds his personal horizons literally expanded with the chance to travel between planets. But technology can also bring disempowerment, and the line between the two is not always obvious. This chapter examines the areas in our lives where technical augmentation is becoming the norm, and reflects on some of the implications.

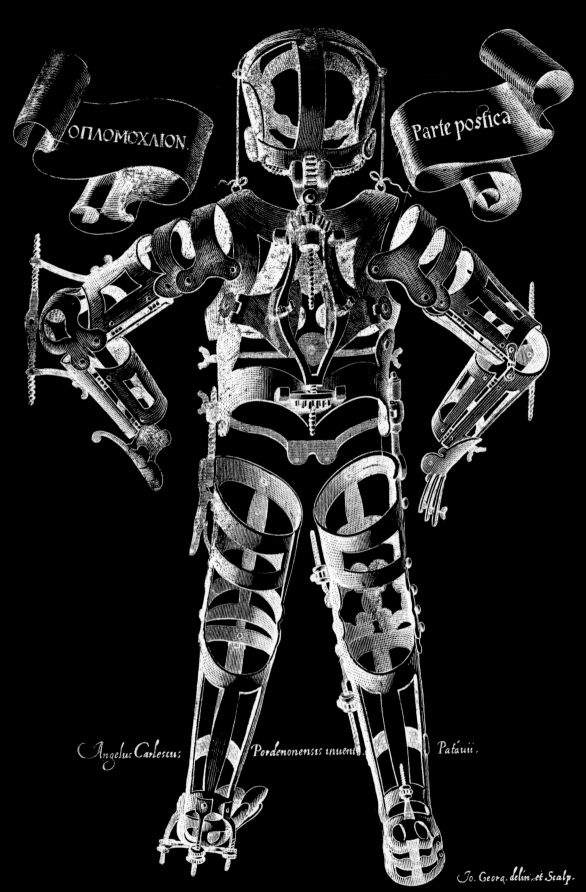

ОПΛОМОХΛIОΝ.

Parte poſtica

Angelus Carlescus    Pordenonensis inuent.    Patauii.

Jo. Georg. delin. et Scalp.

04••
The notion that man is made up
of replaceable parts is shown
in this sixteenth-century
illustration by Hieronymus
Fabricius, an Italian
anatomist and embryologist.
The image appears in his
*Opera Chirurgica*, which
includes a section on surgical
instruments. Medieval armour
was the inspiration for the
exoskeletons – orthopaedic
supporting apparatus –
illustrated in his book.

medical
advances

05●●
The advances in prosthetic design over the past hundred years is due to a combination of greater design awareness and the advent of new materials that are lighter and more malleable. Compare the prosthesis shown below with the device (especially designed for athletics) used by Aimee Mullins on page 101.

06●●
Lower leg amputees can have problems with symmetry of gait. Professor Blake Hannaford and researchers at the Biorobotics Laboratory at the University of Washington are developing this prosthesis, using an artificial muscle and tendon based on a hydraulic-like system. This helps to establish a near-normal gait, and also reduces the effort expended.

07●●
One difficulty for amputees is deciding how much pressure to apply to an object with their artificial hands. Dr Peter Kyberd of the Oxford Orthopaedic Engineering Centre has developed the Leverhume Oxford Southampton myoelectric prosthetic hand. The advanced technology relies on a combination of sensors and computers to give the amputee greater dexterity and control in handling objects.

08●●
Ambroise Paré's compassionate skill in treating wounded soldiers was such that, when captured, he was recognized and his life was spared. A sixteenth-century surgeon had limited tools and no anaesthetic. Amputations had to be done in thirty seconds, with three minutes to complete. Fatalities were high. Paré designed many prostheses, among them this metal hand that moved with a system of springs and catches.

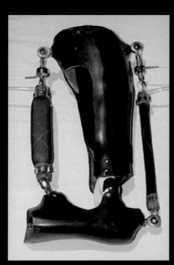
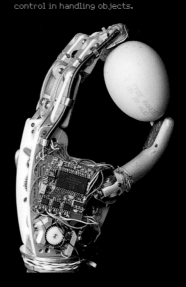
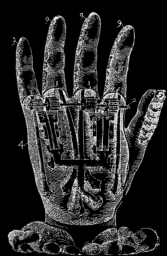

The loss of a limb is traumatic psychologically as well as physically. Historically, prosthetic limbs were able to help compensate for the physical loss. But the various devices failed to give amputees some sense of their missing limb. The connection between the prosthetic and the body was basic, using fixings that offered limited movement. Ambroise Paré (1517–90), a military surgeon, graphically illustrated designs for prosthetic limbs in his 1564 field guide for performing amputations. His designs resembled suits of armour, made from sheet metal with leather straps to keep them in place. The more advanced designs used a hinge system so that the limb could be manually raised and lowered at the joint. Primitive in retrospect, they were considered very advanced in their day.

Not all amputees experience total loss of the missing limb, as some retain the sensation that a limb still exists after amputation. This is known as Phantom Limb Syndrome. Pain and discomfort are the most frequently reported experiences. In 1998, artist Alexa Wright worked with eight amputees to investigate and visualize the specific characteristics of their phantom limbs. Each person represented was interviewed and photographed, and the images were then digitally manipulated under the direction of each individual to gain a true representation of the phantom. One of the surprising comments to emerge from the interviews was that, although all the amputees experienced negative sensations, none would prefer to be without the sensation. The feeling of pain or discomfort somehow served as a reminder of their missing limb, so that the loss did not appear quite so great.

Contemporary prosthetics and implants aim to create a more direct link with the sensory control of the amputee. Scientists are starting to look at ways in which connections might be made between nerve endings and the brain. These could control movement and provide sensation. The origin of this technology dates back to the eighteenth century and the work of the Italian scientist Luigi Galvani. Galvani's work began in 1770 with an accidental observation of the twitching of the legs of a dissected frog. This movement had been caused when an exposed nerve was touched with a steel scalpel, encouraging sparks from an electric machine nearby. This gave rise to Galvani's theory that a nervous electric fluid was secreted by the brain and conducted via nerves before being stored in our muscles. The process later became known as galvanism. Today, electricity is still being used, but in a more controlled way. Professor Ifor Capel and Dr Helen Dorrell have developed a Sub Perception Electro Stimulation (SPES) system. An undetectable electric current is transmitted to counter chronic pain when drugs are no longer effective, and has even been shown to help alleviate certain allergies such as hay fever. SPES is also the Latin word meaning hope.

In England, the Oxford Orthopaedic Engineering Centre has developed a robotic hand that opens in response to myoelectric control signals from the amputee's muscle contractions. The amputee can generate these impulses through brainwaves so that thinking about the action will make the hand respond. The prosthesis follows two motions based on the most common grip postures while a force sensor in the palm gives force feedback so that the amputee no longer has to rely on sight for control. The symmetrical design allows the left and right hands to be made from the same basic set of parts, and they are covered with a silicone rubber glove cast from a real hand to give a realistic appearance.

In 1997 Alexa Wright worked with eight amputees to investigate and visualize the characteristics of their 'phantom limbs'. The images here come from a series called *After Image*. Each person was interviewed, and, under their direction, photographs of them were digitally manipulated to gain a true representation of their phantom.

'I can't imagine being without the phantom because it is there all the time and it is very much like eating or breathing; I can put up with it quite adequately and would probably miss it if it went away.'  RD

'Most of the time the phantom just feels flat; I have to think about it to make it a solid form. I wasn't born like this and obviously I do miss my arm, yet sometimes the phantom pain makes me feel whole again.'  JN

11 ●●

The Italian physicist Luigi Galvani
conducted numerous experiments
on frogs' legs with electricity. His
belief that the muscles themselves
contained electricity resulted in
the theory now known as galvanism.
The process is shown in this late
eighteenth-century illustration.

12 ●●

A defibrillator (this one is made
by Medtronic) is a small device
implanted in the body to detect
and compensate for irregular heart
rhythm. Its electronic memory
stores information, which can then
be retrieved during the patient's
next check-up. A patient is often
unaware of any heartbeat
irregularity until the recorded
information is analysed.

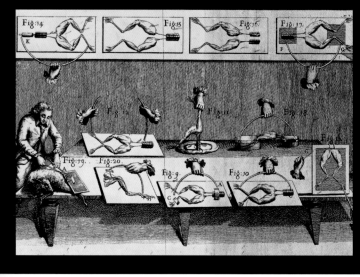

Vision has been amplified for centuries through
the use of eyeglasses, spectacles, telescopes
and binoculars, which effectively made cyborgs
of our ancestors. Nowadays we float lenses
on our eyeballs to correct defective vision,
and also to alter the colour of our irises at will.
Infra-red light improves night vision for both
security guards and soldiers. Virtual Reality has
introduced a range of new technologies that
extend our experience of the world with vision
a central element. Originally developed as a
training aid for pilots, the technology has since
moved into computer games, and most recently
it is being adapted for medical applications.
The University of Washington and Microvision
Inc have developed a Virtual Retinal Display
(VRD) which is designed for military and medical
applications. One benefit may be as an aid for
people with poor vision. The VRD scans a low
power beam of light directly onto the user's
retina rather than a screen. The effect is likened
to painting using three laser sources, red, green
and blue, which combine to provide a full RGB
colour scale.

One characteristic of Parkinson's Disease is a
hesitant gait. However, patients can sometimes
have periods of relief from this at certain stages
of their illness through an effect known as
kinesia paradoxa. A California-based sufferer
of the disease discovered that by following
a track of coins or other similar objects placed
on the floor at stride intervals, a cycle of
movement close to normal walking is instigated.
The patient suggested to the Human Interface
Technology Laboratory at the University of
Washington that spectacles used for watching
TV while moving around might be adapted to
provide this effect. In these spectacles, the TV
image appears suspended in space ahead of
the wearer at just below eye level. By replacing
the TV image with such visual clues as dots,
dashes and other geometric shapes,
Parkinson's sufferers are finding it possible

to move with near-normal gait. The adapted
spectacles are still under development, but early
tests are encouraging.

But there are more extreme medical interventions
for humans than prosthetics. There are implants
that are embedded within the body and operate
on one of two principles. The first approach
simply increases or amplifies a stimulus,
and this technique is used in hearing aids and
cochlear implants that operate by turning up
the volume. The second approach uses a small
electric charge to create links between the brain
and organs. This electric charge, which blocks
out certain brain signals, is one of the most
effective methods used to help control pain (as
with SPES, already described) and body tremors.
In pacemakers and more recently defibrillators,
which are used to help the heart maintain a
regular rhythm, batteries send the electrical
charge through a wire to an electrode placed next
to the wall of the heart. A sensing device in turn

At the HITLab, University of
Washington, 'telesavance' means
the transmission of situation-
awareness by telecommunication.
The Virtual Pilot (ViP) project is
designed to allow two people to
collaborate on a demanding task
with voice and gesture. Here, two
advisers, avatars, are beside the
pilot, and problem-solving and
learning are based on intuitive
communication rather than data.

People with Parkinson's Disease can
find walking increasingly difficult.
This disorder, akinesia, is usually
treated by drugs which can have
side effects. Research at the
HITLab, University of Washington,
uses Virtual Reality (VR) technology
to help patients. VR goggles show
virtual objects and abstract visual
clues to simulate an effect known as
kinesia paradoxia, which seems to
trigger a more normal gait.

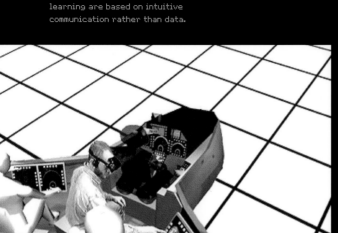

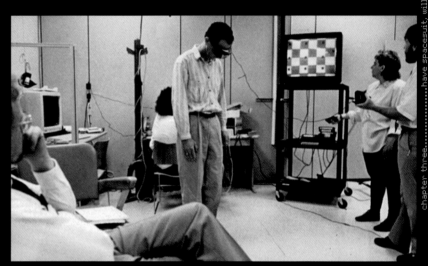

allows the implant to turn itself on and off
as required.

A pioneering form of eye surgery at the Wilmer
Eye Institute in Baltimore offers the possibility
of the restoration of sight to patients who have
gradually become blind. The device used is an
intraocular retinal prosthesis, or IRP. In this
process a microchip is inserted into the retina,
where light patterns are normally converted
into nerve impulses that travel to the brain.
Retina cells that have not been completely
destroyed by the disease are stimulated by
the chip into functioning again. The microchip
transports images using a camera to convert
the external images into a series of electronic
signals. This camera is mounted on to a pair of
spectacle frames that the patient can wear like
eyeglasses. This is a restorative technology,
however; unfortunately IRP cannot be used
for those patients who have been completely
blind since birth.

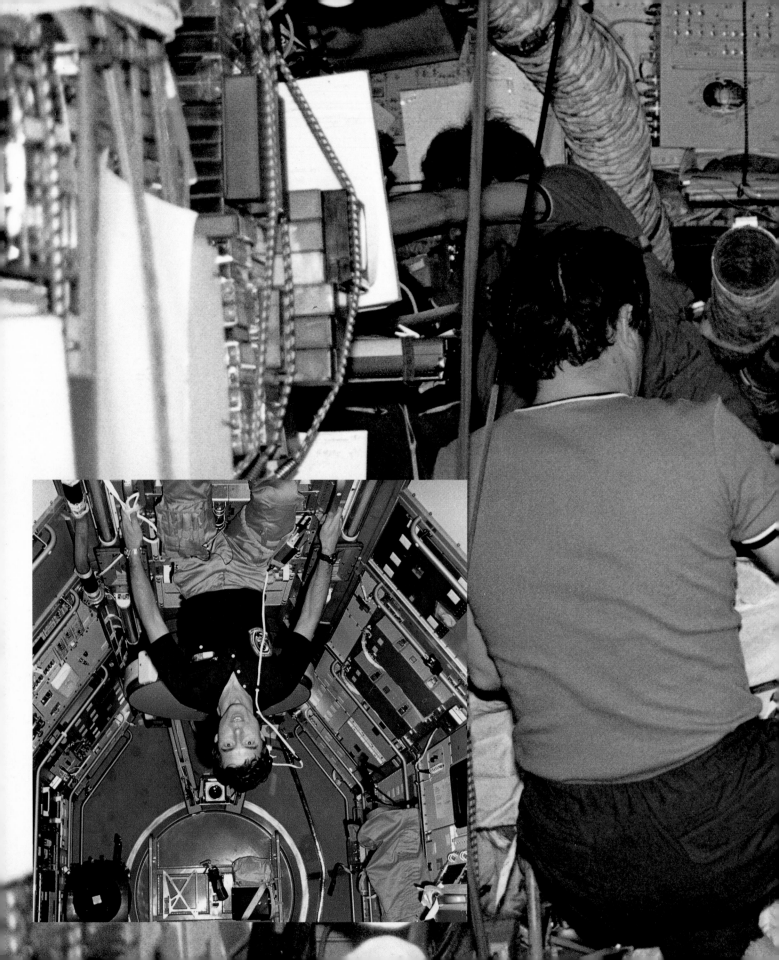

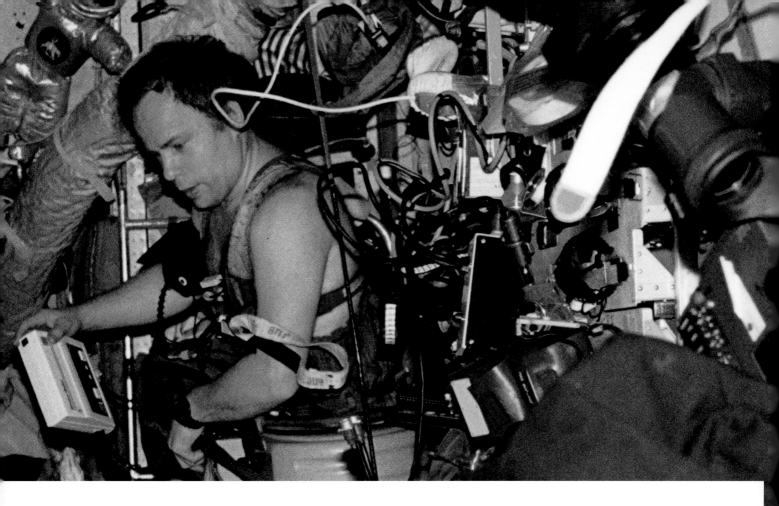

## portable and protective environments

15 + 16●●
While medical implants can be regarded as components of a human body, in the extra-terrestrial environment, the human operates as a component of the machine that is his new environment and without which he would die. The reality of life in space is far removed from Stanley Kubrick's film *2001*. These European Space Agency astronauts must live and work in cramped conditions over prolonged periods of time. Zero gravity causes the astronauts to float, sometimes working upside-down in this weightless environment. More seriously, this atmosphere can cause loss of calcium in the bones and muscle shrinkage, and even the sense of taste is diminished.

Human beings always strive to test the limits of their capabilities. Exploration has provided an opportunity to test human physical and mental endurance, and early explorers of the Earth set forth into risky and uncharted territory. Although the geography of the planet is now well documented, it still proves a considerable challenge, and the most challenging regions of earth are used by scientists to test blueprints for survival in harsh extraterrestrial conditions.

Photographs of early expeditions show men in the most inhospitable climates dressed in tweed jackets and plusfours and other unsuitable clothes. One infamous photograph dates from 1818 and shows a group of British naval officers being greeted by Inuit in Greenland. The officers stand on the packed ice wearing formal regalia while the Inuit are also dressed for a special occasion, but more appropriately for the environment in their parka jackets and boots.

Those who enjoy extreme sports are deliberately courting risks and even death for the thrill of finding out how far they can test their bodies and their nerve. Seven hundred climbers are estimated to have reached the summit of Mount Everest, and a further hundred and forty have died trying. A better knowledge of the terrain and conditions contributes to a constantly improving success rate, as does the use of technologically advanced clothing. Waterproof, breathable fabrics have become standard, while thermo-regulating materials that keep the body at a pleasant temperature have been introduced more recently. The technology for the latter originated with NASA, but it is now being used for expeditions, sport and leisure wear. Mistakes these days are often attributed to poor decision-making and other human errors rather than any failure of equipment. In a 1996 disaster on Mount Everest, eight climbers

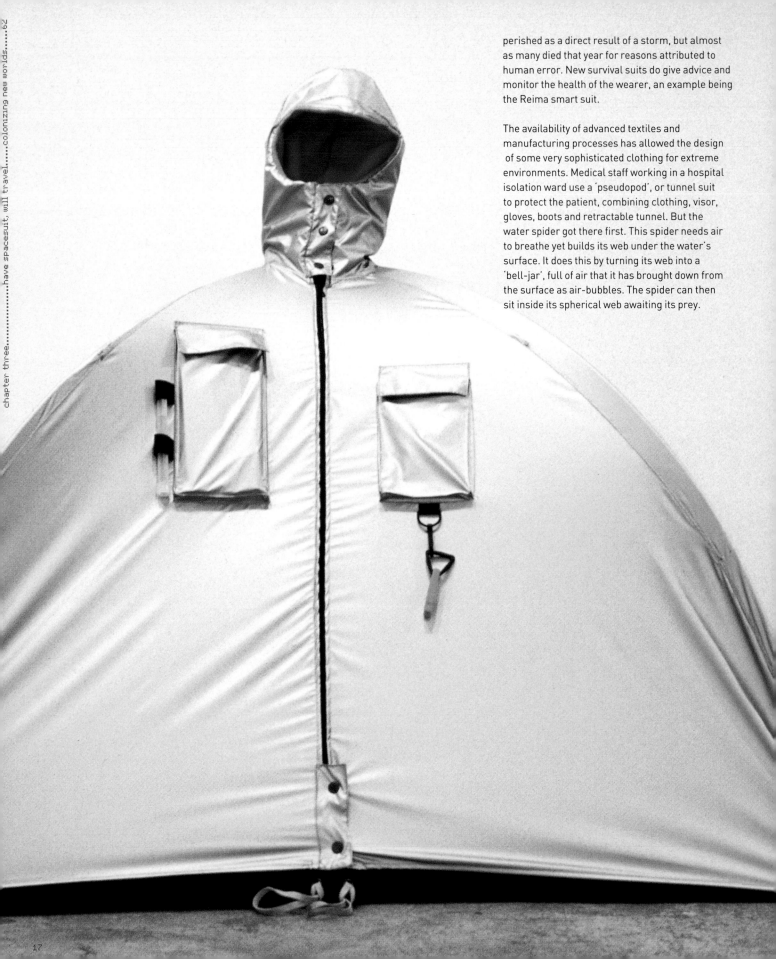

perished as a direct result of a storm, but almost as many died that year for reasons attributed to human error. New survival suits do give advice and monitor the health of the wearer, an example being the Reima smart suit.

The availability of advanced textiles and manufacturing processes has allowed the design of some very sophisticated clothing for extreme environments. Medical staff working in a hospital isolation ward use a 'pseudopod', or tunnel suit to protect the patient, combining clothing, visor, gloves, boots and retractable tunnel. But the water spider got there first. This spider needs air to breathe yet builds its web under the water's surface. It does this by turning its web into a 'bell-jar', full of air that it has brought down from the surface as air-bubbles. The spider can then sit inside its spherical web awaiting its prey.

The paradox of survival in isolated and extreme conditions is that the person is forced to be self-sufficient but also reliant on co-operation with colleagues. Unsurprisingly, such situations give rise to tremendous psychological stress. A study commissioned by the American Navy looked at the importance of human behaviour in an isolated environment. Sociologist Dr Lawrence Palinkas analysed data on long-term isolation from US Navy personnel stationed in Antarctica. He considered external influences such as lack of personal contact with family and friends, as well as stress from real or imagined unpleasant events at home, and studied factors relating to habitat and living conditions, including lack of privacy, cramped quarters, poor stimulation and boredom. Palinkas considered the six months of light and

darkness in Antarctica to be similar to conditions on the lunar surface, which produces a type of culture shock often leading to depression, hostility, sleep disturbance, impaired cognition, as well as physical and psychological distress. It took the subjects about six months to recover once they had left the Antarctic environment.

One theory of how this might be solved suggests that humans should undergo an accelerated learning process to adapt to the new environment. This is partly based on animal experiments where two groups of rabbits were placed in a darkened basket under very different conditions. The first group was carried everywhere, while the second group was tethered to a box that they have to pull with them in order to move about. When eventually exposed to

daylight, the rabbits in the first group were completely disorientated and had difficulty standing. The second group moved about with ease because they were still able to apply their knowledge of motion.

In his book *Living and Working in Space* (1996), Philip Robert Harris describes the effect of extended spaceflights on the astronaut: 'The body gradually acclimatizes to microgravity, and cardiovascular deconditioning sets in, human bones lose calcium, and a demineralization process begins; the space adaptation syndrome of nausea, disorientation or discomfort can be overcome by most spacefarers.' While diet and exercise counteract many of the side effects caused by short-term stays, long-term stays are more problematic.

17 + 18••

The work of British artist Lucy Orta examines the isolation of the individual, not in space or even in the Antarctic, but in our urban environment. The tent-like structure shown is from the *Refuge Wear* series, 1992-94, and includes 'habitent, mobile cocoon, ambulatory survival sac'. The work centres on people living in the streets of cities where the effect of lack of shelter for

prolonged periods can result in a severe deterioration of physical and moral health, increased stress, weakened immune system and loss of identity and desocialization. *Nexus Architecture*, 1997, shows people linked by an umbilical cord-like structure, which serves to emphasize our reliance on others. Humans are not by nature suited to social or physical isolation.

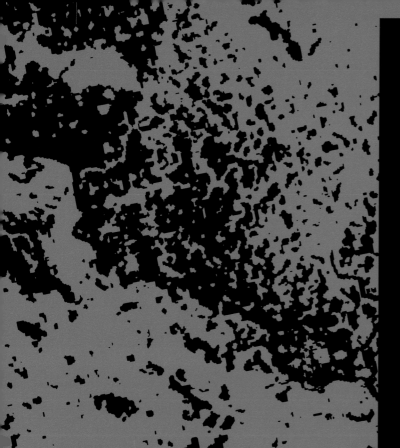

19 + 20 + 21 + 22 ●●

The means through which we watch and are in turn watched are becoming increasingly covert. In the 1954 Alfred Hitchcock thriller *Rear Window* (opposite right), the photojournalist played by James Stewart follows his neighbour's movements using a camera with a highly visible telephoto lens. It is through this that he is eventually discovered and comes face-to-face with the murderer. Today's surveillance can be much more subtle (unless its purpose is as a deterrent) so that few are aware of exactly where global monitoring systems are looking and listening. One purpose of these is to monitor climate changes and record the extent of deforestation in the Amazon (background opposite) or flooding in Bangladesh (this page). Future plans for observation satellites include Envisat (opposite, left) to be launched by the European Space Agency. Equipped with the most advanced range of sensors, it will improve the range and accuracy of scientific measurements of the atmosphere, oceans, land surface and ice.

Further ahead is a vision of an omnipresent surveillance technology so small that it will be invisible to the naked eye. This type of development will be made possible by nanotechnology. In his book *The Diamond Age*, Neal Stephenson explores the possibility of such a world, but, like many science fiction writers, he makes use of existing technological research and development. In Stephenson's vision, clouds of Smart Dust wage aerial guerrilla warfare, sprinkling tiny nanobots across people's shoulders to monitor their every move. A startlingly similar technology (even called Smart Dust) is being developed at the University of California at Berkeley, funded by the American Defense Advanced Research Projects Agency (DARPA). The Smart Dust will consist of autonomous nanobots that can be used to monitor everything, from the quality of the food we eat to the temperature of our environment, and, of course, they will also monitor our actions.

# communication

The irony of communication technologies is that, while bringing people closer who are geographically distant, they can also create a gulf between people in the same room. In a restaurant recently, the author noticed five young women seated at a nearby table. All five were speaking on their mobile phones rather than talking to each other. It was a bizarre scene, but not unusual these days. The media guru Marshall McLuhan predicted our 'global village' more than three decades ago. Is this really what he meant? It is still not known what the long-term effects such technologies will have on people and society. Anecdotally, the signs so far are not very encouraging, and the story of the women in the restaurant is by no means unique.

Sophisticated communications technologies have also dramatically changed the way we view the world around us. Satellites can now see the American flag on the White House lawn. The same satellites can also inform us of changes in our environment, such as deforestation in the Amazon rain forest. And spy cameras can be hidden within false light fittings to monitor the home. In his seminal book *1984*, George Orwell wrote of a society watched over by Big Brother. Rooms equipped with Telescreens showed propaganda but also housed built-in cameras and microphones to spy on people.

One company in the USA supplying home surveillance products is Big Brother Surveillance. The images are sent via the internet to the client's office computer. Its latest products in domestic surveillance could have been taken direct from the pages of Orwell's book. Cameras are embedded in clock radios, VCR machines, coffee makers, smoke detectors and picture frames. They are not just aimed at burglars, but also at childminders whose activities can be monitored from the office by the parent. Information is relayed as binary data that must be further interpreted to be of use. The process of interpretation is usually done by a human and can lack objectivity. Imagine a parent while at work watching a childminder via a home surveillance system. The carer is seen to pull a screaming child across the living room and through a door to another room. What will the parent conclude? Most likely that the child is being abused. But it is also possible that the child has had an accident and is being taken to another room for medical attention. Not everyone will be aware that he or she is being watched or be given the opportunity to defend what may appear to be suspicious.

The word 'surveillance' comes from the Latin term *vigilia*, which means wakefulness or

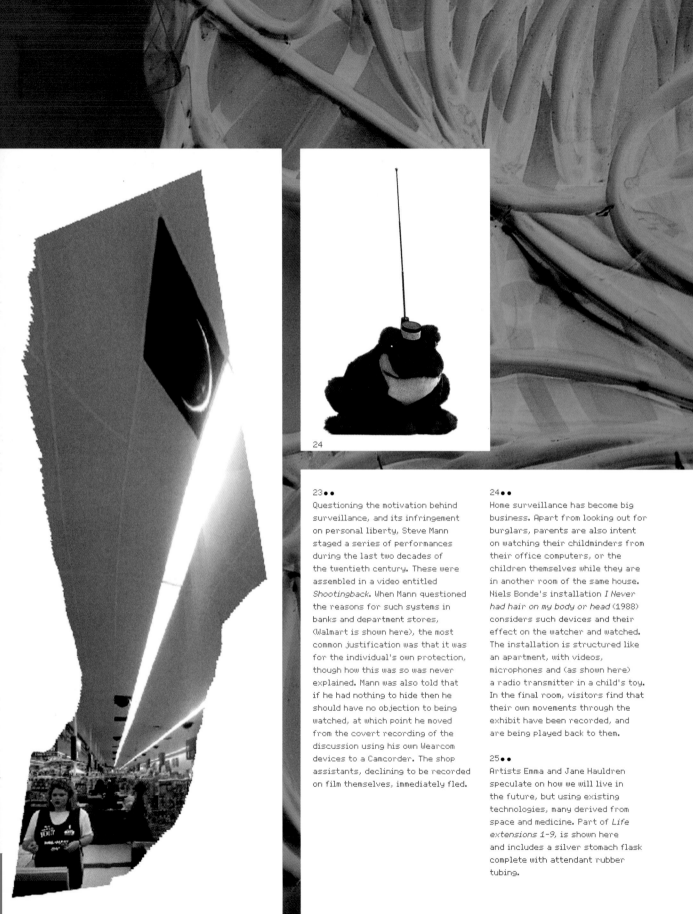

**23••**
Questioning the motivation behind
surveillance, and its infringement
on personal liberty, Steve Mann
staged a series of performances
during the last two decades of
the twentieth century. These were
assembled in a video entitled
*Shootingback*. When Mann questioned
the reasons for such systems in
banks and department stores,
(Walmart is shown here), the most
common justification was that it was
for the individual's own protection,
though how this was so was never
explained. Mann was also told that
if he had nothing to hide then he
should have no objection to being
watched, at which point he moved
from the covert recording of the
discussion using his own Wearcom
devices to a Camcorder. The shop
assistants, declining to be recorded
on film themselves, immediately fled.

**24••**
Home surveillance has become big
business. Apart from looking out for
burglars, parents are also intent
on watching their childminders from
their office computers, or the
children themselves while they are
in another room of the same house.
Niels Bonde's installation *I Never
had hair on my body or head* (1988)
considers such devices and their
effect on the watcher and watched.
The installation is structured like
an apartment, with videos,
microphones and (as shown here)
a radio transmitter in a child's toy.
In the final room, visitors find that
their own movements through the
exhibit have been recorded, and
are being played back to them.

**25••**
Artists Emma and Jane Hauldren
speculate on how we will live in
the future, but using existing
technologies, many derived from
space and medicine. Part of *Life
extensions 1-9*, is shown here
and includes a silver stomach flask
complete with attendant rubber
tubing.

24

23

chapter three............................have spacesuit will travel............colonising new worlds 67

sleeplessness. The implication is that it this is not a restful state of alertness, rather a benign paranoia. The American science fiction writer Philip K. Dick described paranoia as 'a modern day development of an ancient, archaic sense that animals still have – quarry-type animals – that they're being watched.' Dick attributes this to a time when our ancestors were vulnerable to predators: 'This sense tells them they're being watched. And they're being watched probably by something that's going to get them.' The very language used in surveillance denotes a hunter-prey relationship. Celebrities followed by paparazzi often complain that they feel 'hunted' by photographers. But by inviting publicity, the celebrities forego the right to be selective about how or when they are photographed, and as such can be seen to be fair game.

By the year 2000, there were an estimated 150,000 CCTV cameras in London – most of these were located in public spaces. Many are prominently displayed, and such overt surveillance systems are perceived to function

as a deterrent. In many cases a tape is not even running, though people have no way of being sure. Shopping centres and department stores use CCTV cameras to record the activity of staff and customers alike. MIT graduate Steve Mann has questioned the proliferation of this type of surveillance in his video *Shootingback*, 1997. The work is based on a series of staged events in department stores during the 1980s and 1990s. Photography was forbidden in most of the stores visited, although there were a number of CCTV cameras in place.

Computers with human emotions are still some way off. However, at MIT in Cambridge, Massachusetts, the technology has recently been developed to allow computers to recognize emotions in humans. In an essay titled 'Does HAL cry digital tears?', MIT's Dr Rosalind W. Piccard asks 'how can computers become intelligent friendly companions if they are not given at least the ability to recognize such emotions as interest, distress, and pleasure?' A reasonable point, until we consider how such information might be used. The next stage may

be monitoring people in the workplace. Employers could receive an analysis of the information gathered: could they ignore information that indicated that certain employees spent much of their time deep in boredom or in hatred of their job?

'The Iran-contra scandal that once looked deeper and dirtier than Watergate was suddenly transformed by [Oliver] North's performance on network TV into something on the scale of American heroism like Valley Forge or MacArthur's return to the Philippines.' Thus wrote Hunter S. Thompson in the second volume of his *Gonzo Papers, Tales of Shame and Degradation in the 1980s* (1989). The power of television is increasingly pervasive. During the Gulf War (1991), images of the conflict were relayed in real time on Western television screens. Cameras were attached to the interior and exterior of war machines, giving the fighter's view of the conflict. The view from the cockpit was not dissimilar to the view offered by combat video games, making a telespectator of the viewer at home who could feel personally involved.

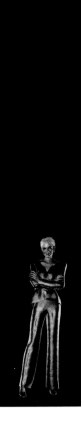

# hyper-reality

Technology now makes it possible for sight, sound, touch, taste and smell to be artificially recreated. Damaged senses can be augmented or existing capabilities enhanced. The technologies can also facilitate immersion in a virtual environment that overrides normal human sensory experience, replacing it digitally.

Although Virtual Reality (VR) provides an immersive environment, the term 'virtual' suggests that the environment will always remain less than the reality it strives to simulate. VR was originally designed to simulate flying conditions for pilots and astronauts undergoing training. Tactile feedback provides a physical sense of reality for VR users. The methods used to provide this sensation include vibrating tactors, single vibrating joysticks and pneumatic gloves. The technology is being adapted for use by deaf and blind people as a means of communication. It is seen as a system that combines object recognition for the deaf and linguistic fingerspelling for the blind.

The computer games industry is probably the most extensive user of VR today, with some of the technologies finding applications in the film and medical industries. The use of technologies inspired by VR in film is certainly enhancing the capabilities of the actors, particularly for special effects. Motion Tracking Sensors are combined with blue-screen montage technologies (which allows two pieces of film to be combined) in the 1999 film *Star Wars: The Phantom Menace*. These technologies bring together the movement of a human actor along with animation to create the character of Jar Jar Binks. The real actor, Ahmed Best, was fitted out with a suit containing reflective markers that are used as motion-tracking sensors, and these are filmed from numerous camera angles. The images are then fed into a computer which calculates the position of each marker, and from this the animated character is created, retaining the actor's distinctive gait and mannerisms. Similarly, in the making of *Toy Story 2*, 1999, actor Tom Hanks was filmed as he read the script. Expressive body and face movements were translated on to the storyboard so that the character of Woody could display some of Hanks's mannerisms.

Ultimately, we may see human actors replaced by their digital counterparts. These cyborgs will never be late on set, put on weight or grow old. Some filmmakers have long dreamt of digitally recreating their favourite actors. In 2000, telecommunications giant Motorola commissioned Digital Domain to create an animated personality as the face of its new voice-activated web browser. The character of Mya is based on actress Michelle Holgate, who was extensively photographed and the images scanned into the computer. Amazingly, the initial result was so lifelike that viewers were unimpressed because they thought it was a real woman. So a revised version was made to look more artificial, with shiny skin actually modelled on the lustre of a china plate.

The individual senses are being enhanced or, in some instances, replaced digitally. A 'nose on a chip', called e-nose, has now been developed by CALTEC using an artificial neural network (ANN). This method of processing information is inspired by the way in which the brain works, although less data is processed. The e-nose is anticipated to have applications in the food and drinks industries, environmental monitoring as well as telesurgery. An electronic nose could examine odours from breath, wounds and body fluids, which could be of use to the doctor as a diagnostic tool. They could provide information on changes in the patient's condition more

quickly than traditional monitoring procedures. However, the e-nose is still quite crude in comparison with the human olfactory receptors – an electronic nose contains a few dozen sensors, while the human equivalent has millions. So, for now, nature is ahead of this particular cyborg sense.

The world and the universe, and our relationship with them, are changing dramatically. The pace is such that our bodies cannot adapt quickly enough and enhancement is becoming a necessity. Technologies developed for space and military applications used to take twenty or thirty years to bring to consumer applications; today a decade is considered a long time-lapse. There is a greater need to be mindful of what these technologies are, how they are being used and by whom. The monitoring of workers really only became a civil liberties issue once the surveillance technology was already installed and in use.

Although the augmentation of the senses and the body is increasing very quickly, there may come a time when it is difficult to distinguish the natural from the synthesized. Ultimately, if the interface between humans and the environment

becomes programmed it will be open to interference. A persuasion perfume for example, could be worn to distort a cyborg sense and influence key decisions. If this programming could take place with all the senses, through cyber-branding, manufacturers could control our entire experience of the world ● ●

26 ● ●
Motorola have introduced Mya on its Myosphere Internet service. The company describes her as 'digitally animated with a personality and look as slick and stylish as the service she offers'. The initial design for Mya was so life-like that viewers thought it was a real woman and were not impressed. The final (revised) shiny version is shown here.

The use of drugs or laboratory techniques to alter humans causes great apprehension. It may not help that, unlike mechanical adaptations, our genetic information is invisible and modifications of it are irreversible. Yet advances are moving apace. Fifty years ago doctors were experimenting with the effects of hallucinogenic drugs on patients. Today, doctors debate the ethics of cloning and the possibility of creating hybrid beings. The danger of altering nature was a central tenet of Greek mythology but is science in danger of outstripping even the most fantastic of these tales?

The gypsy folk rhyme recited in the 1941
film *The Wolf Man* (played by Lon Chaney Jr)
suggests there is little man can do if fate
is against him:

'Even a man who is pure in heart
And says his prayers at night
May become a wolf when the wolf-bane blooms
And the autumn moon is bright.'

Is science creating monsters in our
midst against whom we will be helpless
to defend ourselves?

The Industrial Revolution gave rise to notions of standardization and predictability in the production of goods, which quickly spread to encompass the supply of services. In the rush to quantify what is human and to find cures for major illnesses, these notions of standardization and predictability are now being applied to human beings. The exact nature of diseases is being analysed as never before so that precise cures can be found. At the other end of the spectrum, what is 'normal' is also under scrutiny. How much of who we are is hereditary and how much environmental? Although mostly well intentioned, medical science is in danger of creating a displaced class of people who are not an exact fit for the categories being established. As we move further towards selective breeding, with parents now able to establish the sex of their unborn child, and, in some instances, select their egg donor, the dystopian vision of science fiction draws closer – the baby factories of Aldous Huxley's *Brave New World* (1932) where test-tube clones are bred for different functions; the rule in *Logan's Run* (the film of 1976) that no one lives beyond the age of thirty; the bleak society in Margaret Atwood's *The Handmaid's Tale* (1985), where young women are selected to conceive and bear children on behalf of the sub-fertile wives of the ruling class.

What is rapidly becoming obvious is that human nature is far more complex than we could ever have imagined. The painfully slow development of Artificial Intelligence (discussed in Chapter 2) is an example of this. This may be the redeeming factor in the future development of medicine. Successive scientific breakthroughs highlight the intricate nature of the human being, and the dangers of disturbing the complex balance of such an organism. Alongside the anticipation of the cures that medical science will produce in the near future, there is a growing interest in alternative medicines, such as homeopathy or acupuncture. These alternative treatments tend to address the patient's whole lifestyle and history, with the illness seen as a symptom of a greater malaise rather than the whole problem. This holistic approach treats the individual rather than an illness or disease, and provides healing human contact in a world where people are increasingly isolated. It has been accepted that effective conventional medicine also owes a great deal to a proper dialogue and a good relationship between doctor and patient.

01

02●●
This Coloured Magnetic Resonance
Imaging (MRI) scan shows in section
the brain of a young person
suffering from the fatal disase
new variant Creutzfeldt-Jakob
Disease (nvCJD). The top of the image
shows the front of the head with
the folded cerebrum forming two
hemispheres. The two white areas
seen at the lower centre represent
the thalamus diseased with CJD. It is
detected as a 'bilateral signal
abnormality' on MRI. CJD is a fatal
disease. It destroys nerve cells
and causes brain tissue to become
spongey, with patients suffering
symptoms that include dementia
and muscle contractions.

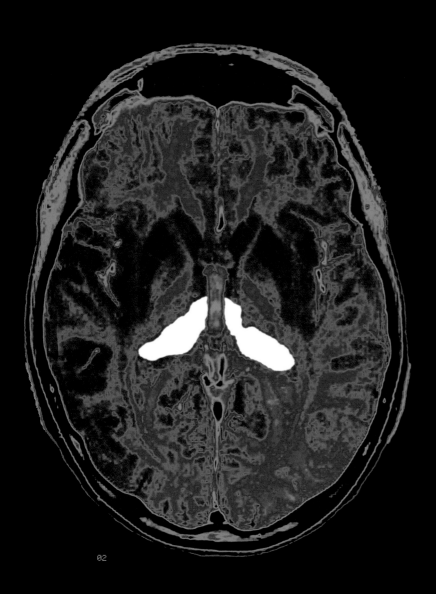

02

## for and against nature

When humans are shown to be out of step with
nature, the moon has often been used as a
leitmotif. The werewolf has come to symbolize
this state of being. Originally a Greek myth, the
werewolf appeared in Ovid's *Metamorphoses*,
written two thousand years ago, where Lykaon
made the mistake of angering Zeus and was
punished by being changed into a wolf. The
werewolf occupies a place in popular folklore
as a tale of horror. Hammer Horror films
invariably depict the character of the werewolf
as a dysfunctional human, undergoing a
metamorphosis at full moon, changing
from human to animal, and embarking on a
murderous rampage that continues until
dawn. It is not until he regains his human
form that he feels remorse for his actions.

Lykaon made the mistake of serving Zeus a
human to eat in the form of umble soup.
Greatly angered by the deception, the god
punished Lykaon by changing him into a wolf.

Angela Carter's werewolf story and the film
based on it, *The Company of Wolves* (1984),
explores the fairytale of Little Red Ridinghood.
Carter's werewolf eats human flesh, effectively
going against nature by eating his own kind.
Also disturbing is Carter's depiction of the
young girl as a willing victim giving herself
freely to the werewolf: 'See! Sweet and sound
she sleeps in granny's bed, between the paws
of the tender wolf.'

The dangers inherent in eating your own kind
are very much in evidence. The outbreak
of Bovine Spongiform Encephalopathy (BSE),
or 'mad cow disease', in the UK, is thought
to have been caused by the farming practice
of feeding cattle carcasses to cattle.
Subsequently, it is believed that some humans

who had eaten infected meat (contaminated by the brain and spinal cord) developed a new variant of Creutzfeldt-Jakob Disease (nvCJD) which causes the brain tissue to develop large holes like those of a sponge. This is not the first time that humans have been subject to this type of illness. Kuru, for example, also a transmissible spongiform encephalopathy (TSE), was diagnosed in 1957 in a New Guinea tribe who practised cannibalism.

The epidemic of BSE and other food scares coincided with a greater awareness of genetically modified foods that were being introduced into the UK. While traditional techniques for genetic manipulation include cross-pollination, selective breeding and irridation, these are now being extended by direct insertion, deletion and modification of genes in the laboratory. Genetically Modified Organisms (GMOs), according to Gentechnikgesetz (Genetic Technique Law) in Germany, are those organisms whose genetic material is modified in a way not found in nature or possible under natural conditions. It must still be classified as a biological unit able to multiply itself and transmit genetic material. On the plus side of the equation, GM holds the promise of cheaper food with a longer shelf-life, and, it is claimed, less need of pesticides and herbicides in growing it. Against this is the fear of what might be unleashed when we interfere with nature. There is concern over possible cross-ertilization with other plants, and fear of an adverse impact on wildlife and on human health, as well as the long-term and unpredictable consequences of introducing entirely new and artificial genetic combinations into the ecosytem. Although GM crops have been grown in quantity in the USA for some

years, in Europe, public opinion is firmly against GM foods and crop trials, and 1998 saw a three-year moratorium on granting licences for the commercial development of GM foods. In February 2001 the European Union moved a step closer to to accepting GM crops when it introduced strict guidlines governing their use. Although the moratorium is still in place, the move was seen by critics as the first step towards lifting it. The new rules include the granting of 10-year permits for growing and labelling GM crops. Meanwhile, in the USA, regulations to stop the label of organic purity being applied to GM crops only took effect in 2001.

The adoption of GMOs have also had far-reaching social and economic consquences. The seeds developed and sold by the biotec companies are expensive. So are the herbicides that have to be used in their cultivation, as the GMOs are specifically bred to be resistant to them. New strains also tend to need more fertilizer and more water than the old varieties. Small farmers in poor countries are finding it more and more difficult to meet all these costs. In 2001 farmers in Brazil were resisting pressure to accept the introduction of GM soya – the soya bean, now comprising around 40 per cent of transgenic crops all over the world, was not previously a food staple for most cultures outside East Asia. The same year saw publication of the Vision 2020 report prepared by the American consultancy firm McKinsey. Disturbing details were published in the *Guardian* newspaper (London, 7 July 2001) which outlined plans for the displacement of millions of poor rural labourers to make way for intensive GM crops to be grown on prairie-style farms in the Indian state of Andhra Pradesh. The pro-GM argument, that such crops are the answer to food shortage in

poorer countries received a further blow when it was revealed that farmers were to be offered incentives to abandon traditional crops and replace them with crops grown for export. Interviewed for the same *Guardian* article, Tom Wakeford of Sussex University's Development Studies Institute said: 'At a time when Britain has put a moratorium on the commercial use of GM crops, it seems hypocritical to endorse their use among some of the poorest people in India.... Nobody is listening to what the poor want.' These questions are relevant to GM animals and humans; GM plants have a longer history, and therefore provide a good indication of the potential ethical, social, physical and economic impact of these technologies over several generations.

03••

In the 1957 science fiction film
*The Fiend without a Face* the evil
thoughts of a scientist (deranged
by an atomic-powered radar
experiment) feeds off the brain and
spinal cord of its human victims.
Invisible for most of the film, when
it does materialize it does so in the
form of a brain using a spinal cord
for mobility.

04••

This seventeenth-century engraving
by M. Greuter depicts a surgery
where all fantasy and follies are
purged and good qualities
prescribed. The shelves in the
surgery (not dissimilar to depictions
of an alchemist's laboratory) are
full of bottles labelled 'modesty',
'honnestete', 'humilite', 'obeissance'
and other 'good' qualities. One
practitioner is shown pouring a
liquid titled 'sagesse' or 'wisdom'
into the mouth of a reluctant

patient, who appears to excrete tiny
jesters into a dish placed under
his chair for this express purpose.
Another, meanwhile, is in the
process of inserting an equally
reluctant patient into an oven.
The follies being purged are
a surreal mix of clothes, musical
instruments, birds and insects –
all seemingly innocuous.

# chemicals

From willow bark extract (an early form of aspirin) to Ecstasy, chemicals have long been used for self-modification. Robert Louis Stevenson's story *The Strange Case of Dr Jekyll and Mr Hyde*, 1886, explored an extreme scenario, in which the use of chemicals became a battle between good and evil. A deadly cocktail of chemicals causes the otherwise mild-mannered Dr Jekyll to break free of any moral compunction, committing heinous crimes as the murderous Mr Hyde. This is perhaps unnerving to consider in the light of new research into drugs to make soldiers into the ultimate killing machines.

The American military are taking inspiration from of one of the nation's favourite comic-book heroes: Captain America. This superhero first appeared in Marvel Comics during World War II. Skinny Steve Rogers is injected with a 'super-soldier serum' to improve his physique and make him fit to fight for his country. His first adversary is none other than Adolf Hitler. In real life, military research suggests that only two per cent of soldiers are capable of continued heavy combat for more than three months. Most of these are diagnosed as pure psychopaths with no conscience or emotional involvement in the killing or dying around them. Military solutions to this problem have varied from alcohol and caffeine to hashish and psychedelic drugs. At one time Timothy Leary, advocate of LSD, was on the Pentagon's payroll as an adviser.

In an essay on 'The Cyborg Soldier', Chris Hables Gray notes that 'it becomes clear that millions of [US] dollars were spent between 1950 and 1975 on the search for drugs that would lower stress and fear while raising or maintaining performance levels.' A SEAL [SEa, Air, Land] team member, speaking about how

drugs were routinely consumed during his time in Vietnam, recalls a sense of heightened awareness and bravado, a feeling of invulnerability. Removed from emotion and conscience, the sole purpose of such experiments is to produce an efficient killing machine. In his book *No More Heroes: Madness and Psychiatry in War* (1988), Richard Gabriel describes the potential of these drug programmes in 'not only keeping soldiers from feeling fear but almost anything else as well, making them functional psychopaths.'

Anthony Burgess considered the ethics of such mind-altering drug programmes in his novel *A Clockwork Orange* (1962). It was written at a time when governments in the West were sanctioning trials into the possibility of using drugs for behaviour control, and this was seen as a possible solution to recidivism. In the book, teenage Alex is introduced with his friends (or 'droogs' in the book's Russian slang) committing various acts of mindless violence. The only hope of reintroducing him to society seems to be a new medical experiment, 'Ludovico's technique'. The system is based on Pavlov's experiments in conditioning dogs, whereby a bell usually announcing food triggers salivation, even when no food appears. Every time Alex even contemplates violence he becomes physically ill. Driven to distraction by the treatment, he attempts suicide. Alex is eventually 'cured' naturally, simply by growing up and tiring of violence.

New treatments for hyperactive or difficult children have their own problems and attendant ethical issues. Children diagnosed with Attention Deficit Hyperactivity Disorder (ADHD) are given a daily course of tablets, such as Ritalin or Equasym, to control their behaviour. The method of assessing the child's

03

04

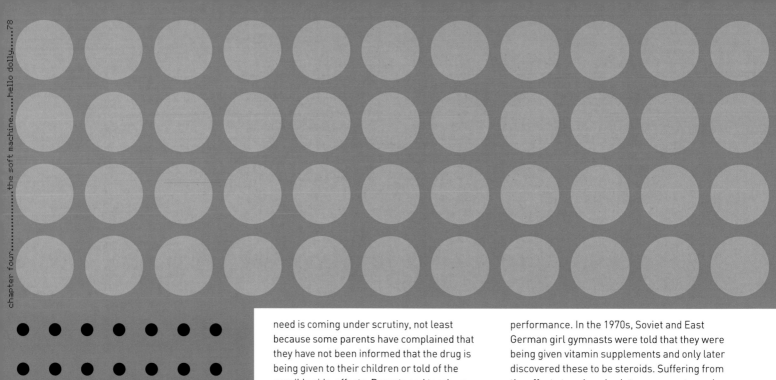

need is coming under scrutiny, not least because some parents have complained that they have not been informed that the drug is being given to their children or told of the possible side-effects. Parents and teachers complain that Ritalin can be addictive.

One mother described her five-year-old as 'possessed, like *The Exorcist*'. Two years later the same child was off the drug and his mother felt him to have matured and become less difficult to manage. Tom Wolfe first encountered Ritalin in 1966 when he was researching for *The Electric Kool-Aid Acid Test*, 1968. He describes a strain of Speed Freak known as the Ritalin Heads: 'You'd see them in the throes of absolute Ritalin raptures.... Not a giggle, not a peep.... They would sit engrossed in anything at all...a manhole cover, their own palm wrinkles...indefinitely.... Pure methylphenidate nirvana.' ('Sorry, but your soul just died', *Independent on Sunday*, 2 February 1997.)

Lauren Slater's 1998 novel *Prozac* describes her physical and psychological response when she stopped taking the prescription drug after ten years. The author describes the shock of finding herself 'normal', feeling assaulted by health, but also confronting the possibility that the drug might not be a reliable or permanent solution to her problems. Her fear is that she is no longer herself without Prozac, and she goes on to question the means for assessing medical success, and the cost of that success to the individual and his or her sense of identity.

Some drugs are taken to alter and control mood and behaviour, and others to improve athletic performance. In the 1970s, Soviet and East German girl gymnasts were told that they were being given vitamin supplements and only later discovered these to be steroids. Suffering from the effects two decades later, some returned their medals in protest. The 2000 Olympic Games in Sydney was notable for the number of failed drug tests and the scandals surrounding them. EPO, or erythopoieten, is the latest rage in performance drugs, especially popular among long-distance runners for its ability to boost endurance.

Performance drugs, however, may become a technology of the past. 'Gene doping will be the next issue,' says Dr Schamasch of the IOC's World Anti-Doping Agency, or WADA. 'Injections of synthetic EPO and human growth-hormone will be nothing by comparison, if somebody increases the production of hormones by direct genetic manipulation.' These vaccines are already being developed for elderly people to increase their muscle mass. They need only be administered once and are undetectable, which will make them open to abuse by athletes and bodybuilders once they become commercially available.

05••
The 1962 film *The Manchurian Candidate* appeared at the height of the Cold War amid general paranoia about Communists in the USA. The film begins with a group of American soldiers, including Raymond (played by Laurence Harvey) being psychologically programmed by their Communist captors. He returns to a hero's welcome, and it emerges that he has a mission to assassinate a Presidential nominee to further his step-father's political career. He has been brainwashed to obey any suggestion whenever he turns up the Queen of Diamonds in a game of patience. The film was banned a year after its release in the USA following the death of President Kennedy, whose assassin was thought by some to have been robotically docile and acting on the orders of others who were never identified.

06••
The Japanese film director Shinya Tsakamoto has achieved cult status with his two Tetsuo films, *Tetsuo: The Iron Man* and its sequel *Tetsuo II: Bodyhammer* (shown here). *Tetsuo II* takes place in a post-industrial world, and the story revolves around the kidnapping of a Tokyo businessman's son. A series of outrages against Taniguchi push him over the edge, transforming him from a mild-mannered businessman to a walking arsenal, half-man and half machine. The more he is angered, the greater the weaponry, until his entire body becomes a rolling tank. As the ultimate fighting machine, he is completely out of control, governed only by his rage and hatred.

05

06

07 ● ●

These two stem cells have been
extracted from blood taken from
the umbilical cord. These stem
cells are totipotent, that is, they
have the ability to undergo
differentiation to form the
precursors to any of the body's
specialized blood cells. The stem
cells develop into red blood cells
or one of three types of white
blood cells that make up the immune
system in a process known as
haemopoiesis.

08 ● ●

In the painting, *She who Spins 1*,
by Jacqueline Morreau, one of
the three Fates, Klotho is shown
spinning the thread at the begining
of a mortal's life. The ancient myth
suggests the immutability of fate.
The question today is whether the
destiny of each individual will one
day be predetermined by medical
science. Will free will become a
thing of the past, or will humans
come to feel absolved of any
wrongdoing because they can
blame their genes?

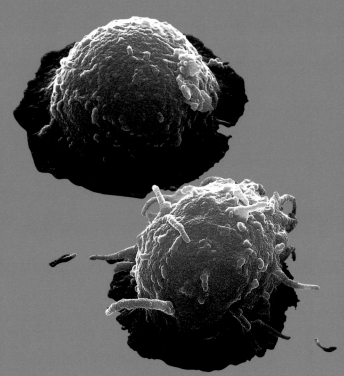

07

08

Limb and organ transplant operations are becoming more commonplace, though both are still being refined. Dr Christian Barnard performed the world's first heart transplant in December 1967, and the following year saw over a hundred heart transplant operations. Heart transplants were greeted with great enthusiasm at the time. They were seen as a daring adventure alongside man's first steps on the moon. The patients themselves had somewhat mixed feelings: 'Seeing my blood outside of my body running through coils of synthetic tubing is deeply distressing... [a] miraculous...powerful monster...with an almost frightening hold on my life... reducing me to a 'half-robot, half-man.' (Fox and Sweeney, 1969.) Limb transplants are proving incredibly difficult. Even when the severed limb is reattached immediately, the body is likely to reject it, or at least create a serious degree of discomfort for the patient. In many cases the patients choose repeat surgery to have the limb removed, replacing it with a prosthesis. Meanwhile, organ transplants are often problematic, and the shortage of human donors has prompted some surgeons to look at alternative sources.

Xenoplantation, the growing of human organs in animals, is one possibility (and is described more fully in Chapter 1). It is a highly controversial area of research, with many people unhappy with the idea of an animal organ, especially the heart, performing a vital function in the human body. In the near future, advances in tissue engineering and stem-cell technology may eventually allow patients to grow their own organ or body part as needed.

Tissue engineers have made significant advances in the creation of neo-organs.

These have the advantage of using the patient's own cell tissue rather than relying on organ donation. This can be achieved in several ways. One option is to inject a molecule (such as a growth factor) into a wound or organ that needs regeneration. A second possibility is to use a three-dimensional scaffolding made of collagen or a biodegradable polymer that will disappear harmlessly once the new organ has been grown around it. There is an application for this technology in the creation of artificial ears. The next stage of development in this area is finding a way of encouraging blood vessels to accept the new organ and supply it with nutrients.

The multipurpose role of stem cells in the body has led to some important advances in tissue engineering and the treatment of brain injuries. In the early stages of an embryo's development, these stem cells (referred to at this stage of development as totipotent stem cells), are self-renewing and have the ability to grow many different parts of the body from muscle to brain cells. Stem cells only begin to specialize four days after fertilization, forming a hollow sphere of cells called a blastocyst, one part of which is a cluster called the inner cell mass, which consists of the stem cells that will go on to form most of the cells and tissues of the human body. These pluripotent stem cells differ from totipotent stem cells in that they do not develop into a complete organism. They are therefore potentially capable of forming only certain types of cells, such as muscle, nerve or blood cells. Scientists believe the pluripotent stem cells are vitally important in gaining a better understanding of the circumstances under which disease-causing aberrations occur and the decision-making process of cell specialization takes place.

Stem cells can be extracted from the umbilical cord at birth. This relatively simple procedure means that the stem cells can be stored and used in the event of the child needing an organ donation or suffering from a debilitating disease. More controversially, however, stem cells can also be taken from the foetal tissue of terminated pregnancies. Influenced by ethical considerations, scientists are now trying to discover whether they can use the more advanced stem cells from children or adults. This is proving difficult. The first step is to change the loyalty of the cells and persuade them to develop differently, and the second step is to ensure that they multiply in their new form. Parents are in some cases choosing to store stem cells at the birth of their children to be kept in liquid nitrogen for periods of around twenty years by commercial companies.

The British company ReNeuron have already had some success in laboratory-developed brain cells whose longevity is controlled by a longevity switch. Kept at a temperature of 33°C (91.4°F) the cells multiply rapidly. Once raised just a few degrees to human body temperature, they stop growing. Trials on animals show that these brain cells can help solve some movement and thinking problems from strokes.

Genes, the vehicle of heredity, are formed by sequences of a molecule called deoxyribose nucleic acid (DNA). When fertilization occurs in animals or plants, two sets of chromosomes (chromosomes are strings of DNA), one from each parent, join to form a pair in the nucleus of the embryonic cell. This cell proceeds to divide and multiply, so that the nucleus of each cell of the living organism contains identical genetic material. The reductionist scientific view used to prevail that the living organism is rather like a machine, with the genes similar to separate nuts and bolts. Genetic engineers even now tend to see genes as if they were as interchangeable as machine parts. On the other hand, non-reductionist scientists believe that genes act in complex relationships with other genes, with other parts of the organism and with the environment. In both scenarios, the mapping of human genes has been the most important breakthrough since the structure of DNA was first identified as a double helix by James Watson and Francis Crick.

Two teams of scientists have worked on the sequencing of the human genome, and ultimately decided to declare their results jointly. The Human Genome Project comprises scientists from governments, institutions and charitable trusts. A commercial company, Celera Genomics, provides the other team. The motive for the research was to gain a better understanding of the human organism, and, in particular, how we might use the information to cure or avoid certain illnesses altogether.

Scientists are setting up trials to treat Alzheimer's by gene therapy – altering the patients' genetic make-up. Scientists acknowledge, however, that implanting genetically altered nerve cells might trigger brain haemorrhages or brain tumours. With one fatality following a gene-therapy trial designed to cure a rare liver disorder, it is being asked whether scientists are under commercial pressure to deliver too much too soon. Treatments for Alzheimer's and possibly some forms of cancer are nevertheless expected shortly. It could still be decades before illnesses like heart disease will be successfully tackled by gene-therapy.

There are profound ethical concerns about the use of DNA information. The 1997 film *Gattaca* provided a timely public reminder of the possible use, or rather misuse, of detailed genetic knowledge. Vincent, played by Ethan Hawke, struggles to become an astronaut. In the futuristic world where babies are genetically engineered to be 'perfect', Vincent is born naturally, and is reduced by his 'imperfections' to membership of a genetic underclass.

There are fears that the new knowledge about our genetic code will lead to an obsession with flaws and a social determinism. Employers and insurance companies could discriminate against individuals who are shown to have certain genes. Pharmaceutical companies are already looking at ways in which they might patent aspects of genetic information while being mindful of civil liberties issues. However, the legal position on both sides of the Atlantic is very complex. The US and UK governments have both issued a joint announcement that the basic sequencing data of the human genome should not be patented. The Wellcome Trust (part of the Human Genome Project) also believes that any gene-related patent should describe the function of the gene and its use, particularly with regard to health, allowing sequencing data to be shared, although it admits that patents are necessary to attract investment for research and development.

There are some unexpected results that have emerged from the sequencing. It has been discovered that humans have far fewer genes than originally supposed – somewhere between thirty and forty thousand (as opposed to the expected one hundred and fifty

thousand), and 97 per cent of this human DNA has no known function. Craig Venter of Celera Genomics described his own surprise: 'In many cases, we have found that humans have exactly the same genes as rats, mice, cats, dogs and even fruit flies.... Really we are just like identical twins, but like all twins and brothers and sisters, we are all really different in the way we respond to the environment.... In everyday language the talk is about a gene for this and a gene for that. We are now finding that that is rarely so. The number of genes that work in that way can almost be counted on your fingers, because we are just not hard-wired in that way.... We simply do not have enough genes for this idea of biological determinism to be right.' (*Observer*, 11 February 2001.) Scientists are now looking at the way in which genes are switched on and manufacture proteins, believing that this causes the significant differences between mammalian species. A key difference may lie in the way in which human genes respond to environmental stimulation.

Behavioural biologists attributed much of the human personality to inherited genes. This included sexuality, alcoholism and even violent crimes. In 1985 a young mother from Georgia was convicted of killing her child, but was later cleared on appeal when she claimed she was a victim of her genes, as she was showing symptoms of Huntington's Disease, an inherited brain disorder that produces serious delusions and uncontrolled movement. Scientists now believe, however, that our environment has a strong influence in forming who we are – nurture as well as nature playing a part. In practice this is not always an easy distinction to make, as most people live with a parent or other relative

Scientists hope that gene-therapy research will provide solutions to many debilitating diseases. This X-ray (autoradiograph) shows an analysis of the structure of DNA where bands of human DNA fragments are seen, prepared using the technique of gel electrophoresis. The structure of the gene is revealed in the black bands, where many bands may comprise one gene. The sequence marked here is part of the research conducted into cancer gene therapy at the Imperial Cancer Research Fund in London. Scientists have successfully 'switched on' melanoma cancer cells with a marker gene. They are now hopeful of using the genetic switch to express an anti-cancer gene that will help the body to destroy melanoma cancer cells.

during their formative years. There have been long-term studies of identical twins, separated from birth, indicating that, besides features, colouring, weight and height, there are unexpected traits that seem to be shared and so are likely to be genetically determined – a distinctive laugh or gesture, life patterns, for instance. Identical twins develop from a single egg and have identical chromosomes, but they also develop in the same womb at the same time. Biological determinism cannot explain the whole complexity of the human organism and its psychology.

Eugenics, the study of the improvement of the human race, especially by breeding, has been one of the greatest causes of concern since the outset of research into the sequencing of the human genome. There were fears that it would fuel discrimination of all kinds. In his book, *Frankenstein's Footsteps* (1998), Jon Turney describes how the eugenics movement, which began in the late nineteenth century, sought 'a general uplift in quality', but 'in practice their attentions were directed downward...eugenic schemes concentrated on eliminating the unfit – on negative eugenics.' It is no doubt to the chagrin of racists that the Human Genome Project has emphasized the similarities of our genetic inheritance.

09

## 10

Dolly is undoubtedly the most famous
sheep in the world. Born in 1997,
she is the first sheep to have been
cloned from an adult sheep cell.
The research was conducted by
Dr Ian Wilmut at the Roslin Institute
near Edinburgh. The process
involved removing the cell nucleus
from the egg cell of a Scottish
Blackface ewe, which was then
injected with the cultured adult cell
taken from the udder of a six-year-
old Finn Dorset ewe. The udder cell
and egg cytoplasm were electrically
fused, and the egg stimulated to
grow into an embryo in the womb
of a Scottish Blackface sheep.

# clones
# and
# hybrids

Genetically identical organisms, like identical
twins, can be artificially created by cloning.
This involves the transfer of the complete
genetic material from the nucleus of a
cultured donor cell to a mature recipient egg
whose own nucleus has been removed. This
is done using nuclear transfer technology.
Cloning, if it is publicly accepted, will have
an enormous impact on all aspects of nature,
from fruit and vegetables to animals, including
human beings. For the purposes of medical
research, it is argued that it would provide
identical organisms for testing, and lead
to more accurate and rapid results.
Few, however have even attempted any
justification for human cloning.

One exception is a religious movement, the
Canadian-based Raelians, who regard human
cloning as a central tenet. The group was
founded after an alleged encounter with
an alien by their leader, Claude Vorilhon
(a former French racing driver). Vorilhon
claims to have witnessed the landing of
a flying saucer in December 1973 in Clermont
Ferrand in France. The Raelians believe that
DNA is in fact the soul and that reincarnation
can only happen through science – cloning.
The Raelians are actively pursuing this aim
and have set up the first company, Clonaid,

11    12    13

which offers to clone human beings. Clonaid was set up in the Bahamas but cancelled by the Bahamian government, possibly under pressure from French media attention. The organization has since set up another laboratory in a secret location. Human cloning has been banned in Europe and Japan, but legislation is not yet in place in other countries, such as Korea where a Clonaid's website is located. The company also offer a service to Insureaclone. For $50,000 they offer to sample and store cells from a human being for later use. Many who have already approached the Raelians and Clonaid are the recently bereaved, wanting to clone a dead child or other relative. Without the same history or memories as the deceased person, however, the cloned human cannot possibly be exactly the same. If, at some future date it is possible to download a person's brain complete with memories (as in the film *Total Recall*, 1990, made from Philip K. Dick's 1966 story 'We Can Remember it for you Wholesale'), this may be more feasible.

One obstacle at present is the low success rate in cloning animals. Attempts to clone mammals from single-cell somatic (non-reproductive) cells continue to show very high rates of abnormality and fatality. Interviewed in the *Observer* newspaper in 2001, George Seidel of the Colorado State University stated that 'there is an abnormality rate of maybe 30 per cent in cloned animals. In human babies, the normal rate of congenital defects is about 2 per cent, and we wouldn't tolerate a jump to 3 per cent.' The rate is unacceptable to many in the case of animals, and the technology for human experiments is banned in many countries. At present, even cloned plants are shown to develop substantial developmental and morphological irregularities. This highlights another concern relating to the side effects and the impossibility of predicting genetic mutations that could evolve in future generations. Genetically modified pigs and Dolly the sheep, for example, develop premature arthritis.

An Australian biotec company, Amrad, has been granted a 'chimeric' patent by the European Patent Office, which covers the use of embryos containing cells of mice, sheep, pigs, fish, and also human cells. The company denies a wish to create hybrid beings, and claims that the aim is to produce genetically engineered mice for research purposes. Greenpeace fears the possibility that the patented technology will be used to grow human organs in animals for transplants, and has called on the European Patent Office to withdraw the patent, stating, 'the chimeras may be non-human but they may contain human organs, body parts, nerve cells and even genetic codes.'

Mikhail Bulgakov's satire on the Russian Revolution and Soviet society *The Heart of the Dog* was written in 1925 but not published in the former Soviet Union until 1987. The story tells of a stray dog being rescued by an apparently benevolent medical professor. The dog is subjected to an experiment by the professor and his assistant, who implant the testicles and pituitary gland of a dead criminal into the dog's body. The result is a gradual change to half-beast, half-man – an animal in human form. He proceeds to turn his creator's life into a nightmare until the professor manages to reverse the procedure.

There is an abundance of hybrid creatures in mythology. Some are allegorical and serve as a warning, while others are based on human deformities or illnesses. In the chapter 'Images of Bodily Transformations' in *Medical History*, Sarah Bakewell of the Wellcome Trust suggests that a scaly creature known as the 'bishop fish' discovered in Poland in 1531 may have been a hooded seal or *cystophora cristana*. The 'fish boy' of Naples is usually regarded as human, suffering from the skin disease ichthyosis, which gives a scaly appearance. Pliny the Elder brings together countless tales of hearsay in his compendium *Natural History* (AD 77) He writes of the 'forest-dwellers...who have their feet turned back behind their legs' allowing them to run at great speed. The Monocoli are less endowed, equipped as they are with only one leg, but they still manage to hop around with some speed and 'when the

11••
This photograph from 1889 shows Joseph Carey Merrick, otherwise known as the Elephant Man, the year before his death at the age of twenty-eight. He was believed to suffer from elephantiasis, a disorder of the lymphatic system, that causes parts of the body to swell to grotesque proportions. Merrick was forced to earn a living by exhibiting himself in a variety of sideshows, where he was billed as The Elephant Man.

12••
Contemporary fashion designers devise garments that give the appearance of repositioning the parts of the human body. Walter van Beirendonck in his No References collection (Winter 1999/2000) creates this effect with deceptively simple shapes, such as this circle-shaped jacket that can be twisted to store itself in its own pocket. The raised shoulders of the jacket make the model's head seem lowered, creating an unconventional symmetry. Some similarities can be seen in Pliny's 'headless man'.

13••
The great explorers of the Middle Ages brought back stories of the fantastic creatures that they had encountered on their journeys. Marco Polo's written description of the peoples he encountered during his travels in Siberia was interpreted by the illustrator of the early fifteenth-century Livre des Merveilles as variously, a headless man with his head between his shoulders, a man with one oversized foot and a one-eyed giant. This fantastical illustration of 'Three Inhabitants of Siberia' makes liberal use of existing marvels and myths.

14••
Many of the medieval bestiaries were based on human disfigurements. Some ailments were attributed to the 'wrath of God', while ducks were thought to cause webbed feet and strawberry birthmarks. This illustration from Fortunio Liceti's seventeenth-century manuscript De Monstris shows three humans with an assortment of animal, bird or fish attributes. The basis for some of the imagery is likely to be medical. The scale-like skin on the third figure, for instance, is not unlike the appearance of a skin disease known as ichthyosis.

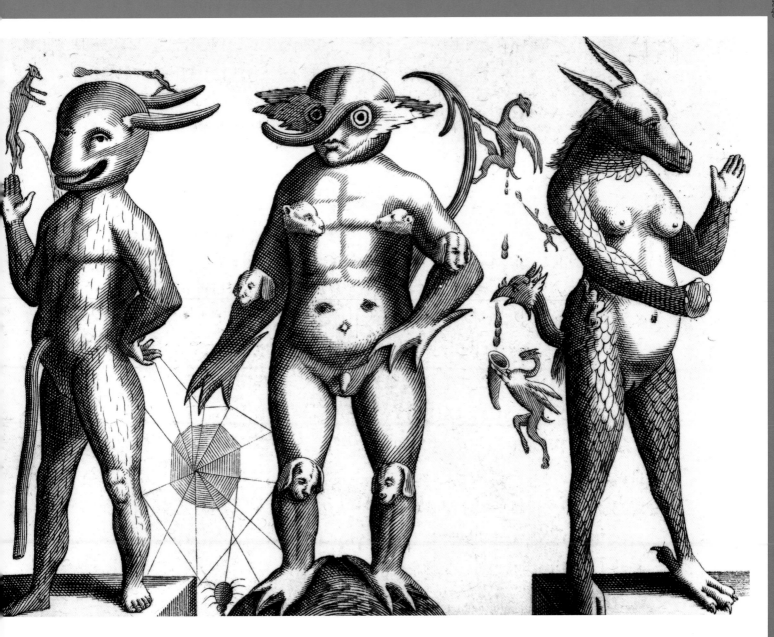

15

16

15●●
The decline of the panda, now an
endangered species, has been
attributed to the clearance of its
natural habitat and its staple diet
of bamboo. Other species, however,
have contributed to their own
extinction. Mauritius was not only
the home of the dodo, but also of
the Mauritius red hen, which was
irresistibly drawn to the colour
red. In an unlucky twist of fate,
seventeenth-century sailor's caps
were made of a standard red cloth.
A sailor writing in 1633 noted:
'They bee very good Meat, and are
also Cloven Footed, soe they can
Neyther Fly nor Swymme.'

16●●
Unusual hybrids are not confined to
mythology or scientific experimentation
– nature can sometimes provide its own
unexpected hybrid. At the Eden Ostrich
World in Penrith, Cumbria, a Shetland
pony arrived and shortly gave birth,
in June 2001, to a zebra-pony hybrid.
The parents had previously been kept
in the same field because mating between
a zebra and a pony was not considered
likely. This type of hybrid is extremely
rare, as horses have sixty-four
chromosomes and the zebra only
forty-four.

weather is hot they lie on their backs stretched out on the ground and protect themselves by the shade of their feet.'

Scientists see biotechnology as a possible solution to endangered, or even extinct species. Pandas are notoriously difficult to breed but the San Diego Zoo successfully resorted to artificial insemination to breed pandas in its care. No doubt inspired by the Spielberg film *Jurassic Park*, 1993, consideration is also been given to the recovery of extinct species. The DNA of long-extinct animals is thought to be too damaged, but it may be possible to recover animals who have become extinct more recently, such as the bucardo, a variety of mountain goat. Tissue from just one bucardo has been preserved, a female. Advanced Cell Technology (ACT) plans to remove one copy of the X-chromosome from one of the female cells and replace it with a Y-chromosome from a goat – the two are closely related. The process would necessitate a surrogate mother, which could possibly be a goat. Greek mythology warns of inter-species union in the story of Pasiphae – Poseidon, in an act of vengeance, made her fall in love with a white bull and she gave birth to the Minotaur, a monster with a bull's head and human body.

In the midst of such remarkable advances in technology, a balance sheet must be maintained. Relieving human suffering is laudable, but account must be taken, as part of the moral considerations, of the cost in animal suffering. In 2001, PPL Therapeutics admitted that in all of its experiments, about 50 percent of all animals are born with abnormalities. Dolly, for instance, was the only successful sheep to be born in ten pregnancies. The procedures do not tend to produce three-headed monsters – most of the defects are kidney and other internal malfunctions. The Roslin Institute in Scotland pioneered transgenic technology for producing greater quantities of therapeutic and nutritional supplements in the sheep's milk. The technology was licensed and patents assigned to PPL Therapeutics, which plans to breed sheep whose milk can be given to humans to treat a deficiency of protein in the lungs as well as cystic fibrosis. Many of these transgenic sheep are born with abnormalities, and the male lambs are incinerated at birth since in the UK genetically modified animals cannot be reared for food. Ron James, managing director of PPL, has no doubt about the justification of such practices, stating, 'You have to look at whether the end justifies the means.' (Interview in the *Guardian*, London, 6 January 2001). This presumes that cures for such illnesses cannot be found using other methods. The reality is that this particular means has cost and volume benefits.

The philosophical and ethical issues tend to be evaluated in human terms, and the plight of laboratory animals is all too often left to a small number of high-profile action groups who counter violence with violence. All animals feel pain, and higher primates experience sensation much in the same way as humans – that is largely why they are used in medical experiments. Animals are not always analagous to humans in their response to chemicals. Famously, had penicillin been tested on guinea pigs it would never have been approved. Penicillin is toxic to guinea pigs, though not to mice or rats, and is of great benefit to most humans. As advances in genetics and cloning proceed, safeguards need to be put in place to protect these animals.

But it may not only be animals that need to be safeguarded against scientific excesses, as the 1932 film *The Island of Lost Souls* shows. The film is based on H. G. Wells's 1896 novel *The Island of Dr Moreau*. The doctor is essentially a crazed vivisectionist, driven to an island retreat by objections from animal-rights activists in the USA. His animals are half-human, and controlled by painful implants. One appears so human that she is taken by the narrator to be the doctor's own daughter, though she is wholly reliant on a serum to keep her human appearance. While science does produce the occasional 'Dr Moreau', most scientists are responsible and aware of broader social issues. But they cannot make the moral decisions for society – as the people who will benefit from the products derived from these technologies, we, too, must accept our share of the responsibility by not ignoring the manner in which they are produced●●

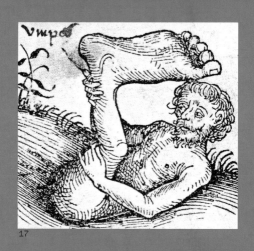

17

**17 + 18●●**

The medieval bestiaries were primarily aimed at adult audiences, however, similar imagery is often used in today's children's cartoon characters and toys to very different effect. While the first were intended to fascinate and horrify, the second provide pure entertainment. Rumpus's brightly coloured Science Freak range of soft toys are aimed at children. One character is Cyclops-like, with one eye, while another has its basis in the sciapod, an example of which is shown (left) from Hartmann Schedel's late fifteenth-century *Liber Cronicarum*. The Dr Seuss *Cat in the Hat books* also feature a range of unique beings, such as the sneetches and the grinch, that seem to morph human and animal characteristics, while Maurice Sendak in *Where the Wild Things Are*, shows creatures with large eyes and mismatched body parts – a lion's body is combined with a horse's tail and scaly legs.

18

The Industrial Revolution saw the creation of a machine environment. Our attitude to machines has varied from resistance, to resignation, and then acceptance. However, there is a new, more human-centred approach emerging in the design of machines that may really improve the way in which we live and work with machines.

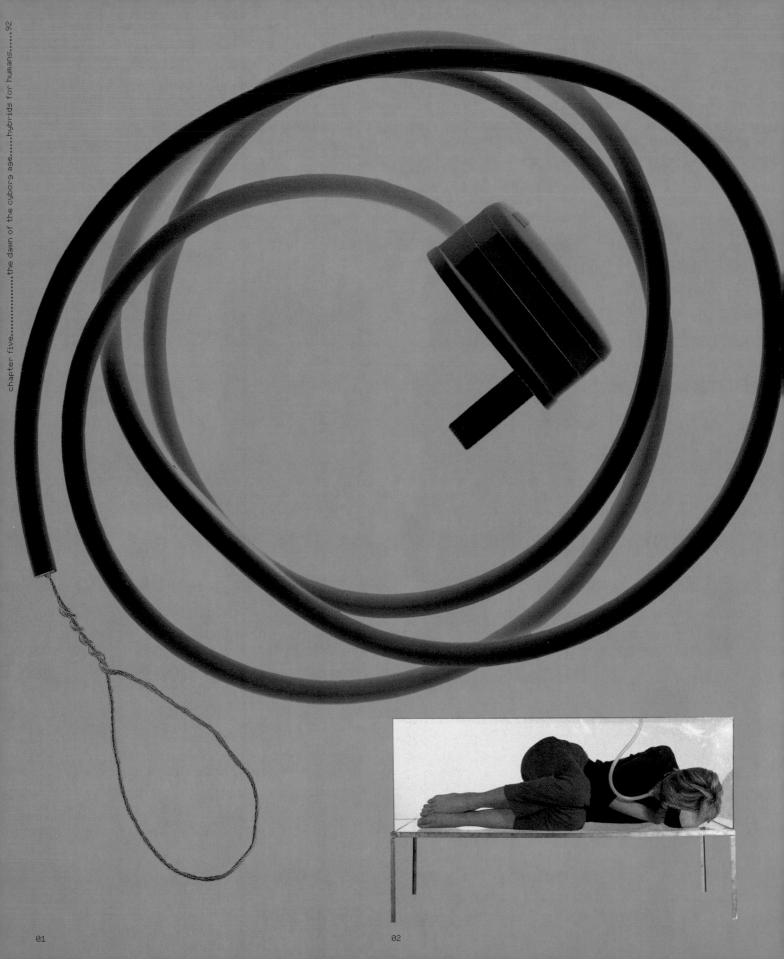

**01••**

Not everyone is reassured that our electronic environment is entirely safe. 'Shots' of wheatgrass are thought to have a beneficial effect on anyone working with computers over prolonged periods, while others are convinced that electricity seeps into their bodies and remains there. Tony Dunn and Fiona Raby's *Plug In* device is based on reports of people who tie a piece of string around one of their fingers and link it to a plug going into a wall socket, believing that this will induce electricity to leave their bodies and return to the National Grid.

**02••**

In *Faraday Chair*, Tony Dunn and Fiona Raby explore a space empty of electronic and radiowaves. (The protective Faraday Cage uses metal, copper in particular, to guide any electric discharge safely to the ground.) Invisible to the eye, the work raises the question, if the inside is empty, what is outside? The human finding solace in the 'empty' space has confined herself to a claustrophobic cocoon and a paranoia more damaging than the electronic environment outside.

**03••**

In *The Ecstasy of Communication* the philosopher Jean Baudrillard complains of the lack of new aesthetic experience through electronics. In a department store in Kyoto, the Body, Mind Health Environment Capsule offers shoppers a new experience somewhere between that of a flotation tank, massage and aromatherapy. All is conducted under the gaze of fellow shoppers.

**04••**

Not all machines are intended to bestow positive benefits on humans. British artist Simon Costin's *Choker* is designed for a single use only. Immediately it has been fastened around the neck, it begins to contract, eventually strangling the wearer. Once closed it cannot be removed.

---

Our relationship with the machine environment has been extensively chronicled over the last century in literature, cinema and the visual arts. This has provided a useful barometer of public attitude towards the machine and the influence it has on our lives. The Italian Futurist movement of the early 1900s, for example, enthusiastically embraced what was seen as the Machine Age. Balla and Depero, two of the leading proponents of the movement, announced the arrival of a metallic animal that could speak, shout and dance automatically. The Futurist view of the world was that it was a force in a constant state of flux. In his manifesto, the poet and art critic Filippo Marinetti stated that the Futurist movement's objective was to celebrate the 'glories of speed, physical aggressiveness, war and the modern machine.' One means of expression used by the Futurists was photography, a relatively new technology at the time. Ironically, the effect of stills photography is to freeze moments of life rather than imbue the subject with any sense of dynamism. The Bragaglia brothers were conscious of the limitation of the medium at that time, and they set about developing what they termed 'photodynamism' – attempting to portray movement in stills photography. The brothers' work, entitled *Polyphysiognomical Portrait of Boccioni*, which was executed in 1913, was a composite photographic portrait of a subject who appears to face in four different directions. The eyes are aligned, but the ears and nose overlap and double-up. When viewed from a distance, this photograph takes on the appearance of a modern hologram.

The Constructivist El Lissitzky, urged fellow Soviet artists towards the greater use of ruler and compass as being more appropriate tools for artists working in the machine age. In 1920, he sketched his vision of a new improved man as he set about redesigning humanity.

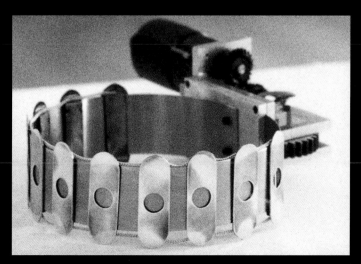

05 + 06••
The design of the Swiss architects Herzog and De Meuron for two signal boxes in Basel makes visible the (electronic) purpose of the buildings. The signal boxes are situated next to railway lines, and house electronic equipment. The environment within the structure is rendered safe, but the architects have chosen to give the exterior a louvred layer of protective copper for aesthetic rather than practical reasons. The signal boxes are reminiscent of the protective Faraday Cage.

07••
Charleroi/Danses explores the relationship between dance and architecture with leading architects such as Zaha Hadid (with technologists including the company 6 Consultants). *Metapolis Project 972* juxtaposes the human with a real and virtual urban environment – the latter through the use of blue screen projection. The resulting dynamic makes it impossible to draw the different elements apart as they come together to read as a single entity of man and machine city.

08••
The sixteenth-century philosopher René Descartes suggested the possibility that a malicious demon had created the illusion of an external reality through control of our senses. The immersive technology of Virtual Reality allows us to create our own illusions in a customized artificial reality. In the film *The Matrix*, 1999, humankind is enslaved in a Virtual Reality created by Artificial Intelligence. The film's hero Neo (Keanu Reeves) has to save humanity from the complacency engendered by this artificial world.

09••
Fritz Lang's architectural training had considerable influence on the film set of *Metropolis*, 1927. The original design showed a Gothic cathedral in the midst of a machine environment, symbolic of the contrast of humanity and mechanical harshness. Lang wanted to focus on the darkness of industry without respite, and wrote on the drawing, 'Away with the church; Tower of Babel instead.' In this scene the human Maria is disempowered while the robot dominates the screen.

Aided by the new tools available to him, the new geometric man is depicted by El Lissitzky in angular, dynamic movement striding across the page. His brain is ideologically programmed with stars, literally, where his eyes should be.

In general, the medium of film has taken a more negative view of the machine environment. Fritz Lang's film *Metropolis*, made in 1927, depicts a world where the human is literally consumed by the scale and power of the machine, having little value other than to maintain the machine. Charlie Chaplin mocks the relentless factory conveyor belt that reduces men to automatons in his 1936 film *Modern Times*, and eventually joins the machine in an apotheosis of shorting circuits. This is obviously not a world that people then or now care to embrace. Such a dystopian view continues throughout the later history of cinema in films such as *Bladerunner* and *The Matrix* (1999).

As prosperity increased after the Second World War, however, there was a general feeling of optimism towards technological advance. This was reflected in lifestyle magazines which in turn were memorably satirized in Jacques Tati's classic French film *Mon Oncle* (1958). This film sees the uncle (Monsieur Hulot), struggle with the conveniences of his sister's house which is packed with every conceivable gadget. This abundance of technology conspires to make life more difficult as human lives revolve around the requirements of the technology. Today's consumer increasingly demands the design of human-centred machines that enable and encourage interaction. Yet the designs are still often lagging behind such aspirations in terms of ease of use.

08

09

10••
The development of conductive
fabric is vitally important to the
design of wearable computers if
they are to be properly integrated
into everyday clothing. This early
prototype for such a fabric was
developed by Maggie Orth and Rehmi
Post at the MIT Media Laboratory.
The woven fabric uses a silk warp
with a weft also of silk but wrapped
in a conductive copper foil. It was
shown in this form at the First
International Symposium on Wearable
Computers held at MIT in 1997.

11••
Wearable computers with head-up
displays are being designed
following very different principles
from conventional desk or laptop
computers. Designs such as Vu-man
developed by Carnegie Mellon
University are conscious of the
need to allow the user as much
freedom of movement as possible.
Unlike the conventional role of the
computer, in this environment it is
generally the user's secondary
activity so its demand on vision
and hands should be minimal.

## human-centred machines

In a consumer society, goods are replaced
many times in a person's lifetime.
Electronic products in particular have
a built-in obsolescence. The public can choose
their purchase from a wide selection, and this
serves to focus the minds of manufacturers
and designers on the creation of more
human-centred machines.

The Wearable Technology Group at MIT's Media
Lab attracts interest from a diverse group of
sponsors which reads like a *Who's Who* of the
electronics industry. As a result, it has brought
together some of these different (and not so
different) concerns to bear on cyborg
developments. Speaking with delegates at
the First International Symposium on Wearable
Computers (ISWC) in 1997, the author found
little attention being paid to the human
element, specifically ergonomic issues such
as the overall weight, weight distribution and its
effect on neck, shoulder and back posture. This

12••
When George Eastman designed the first 'portable' camera he was reportedly asked if he was going camping because of the amount of equipment he had to carry in order to take a photograph. This photograph shows a group of researchers from the MIT wearable technology group. As recently as the mid-1990s the technology for wearables was too bulky, uncomfortable and heavy for most people to consider using. By the end of the century, a combination of commercial forces, awareness of ergonomic factors and the pace of technological miniaturization had paved the way for more discreet user-friendly devices.

13••
One issue with wearable computers is the amount of power needed to operate them. This early Wearcom wearable computer from Steve Mann has its power supply and operating system (a Pentium II processor) stored in the backpack. It was exhibited as shown as here at the Stedelijk Museum, Amsterdam, in 'The Soft Machine: Design in the Cyborg Age', 1998/9.

---

was somewhat surprising given that some of the students were already wearing their head-up display systems for much of their daily lives.

Historically, one of the most dominant influences in the development of wearable technology has been the military. This was evident at the ISWC in 1997. More commercial interests, however, are now bringing human factors to the forefront of development criteria. The effect of electronic extensions on the human body is an area that such artists as Laurie Anderson have been concerned with for some time. In *Drum Dance* for *Home of the Brave* (1985), she designed a suit that had built-in electronic drum kits. The artist describes it as 'the ultimate portable instrument.' The performance piece came about when she was trying to repair a cheap electric drum kit. She discovered that the sensors still worked when the cables were some distance from the circuit board.

The sound generated by the drum suit was so loud that Anderson was forced to choreograph the performance around it, making the dance 'bigger, wilder.'

Thad Starner is a graduate of the Wearable Technology Group at MIT and now Chief Technology Officer at Charmed Technology Inc. The company is a spin-off of the Media Lab, and was set up to produce products and services relating to internet use (e-mail, etc.) that are both wearable and affordable. Other partners on the project include the University of Rochester Center for Future Health in New York. The company has introduced a conceptual prototype and actual products on to the market. The products are noticeably more discreet and 'designed' in comparison with those worn by Starner and his colleagues at MIT just four years earlier. The new designs benefit no doubt from the ever increasing miniaturization of technology, but they also

recognize that for many users the computer will be of secondary importance to another activity. While the person working on a PC or laptop focuses full attention on the computer, the user of the wearable computer will almost inevitably be involved in at least one other activity – walking, jogging or speaking to another person face to face.

Reconciling remote and physical communication is a major challenge for companies like Charmed Technology. Their design (in collaboration with Georgia Tech) for a Hands Free Food Inspection System uses what are effectively goggles as a headset, leaving the hands free to examine food. This makes it difficult to have any eye contact with co-workers, and consequently also communication. In contrast, the Charmed Communicator, which is designed for more general use, has a single screen. This is placed away from the eye to allow eye contact to be

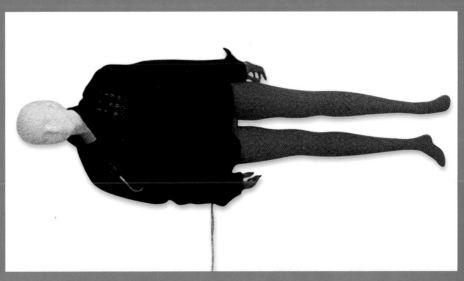

made with other people. The overall design leans towards the computer as fashion accessory, and is based on existing products, such as spectacles, belts and jewellery.

At MIT's Media Lab some scientists are dismissive of wearables as a consumer product. In *Time* magazine (4 June 2001), Steven Schwartz asks, 'Why would you want to surf the Net or play a computer and walk around?' His colleague Richard DeVaul adds that 'a wearable has to be aware of where you are, what you're doing – and give you information accordingly.' This approach is already being adapted by the wearable group at Carnegie Mellon where they are concentrating on the functional requirements of wearable products using feedback from existing users. They have identified the paramount user issues as being energy requirements and supply, as well as flexibility within the system. They are in the process of developing a Spot Core Module based on these principles. The wearable computer now has an external power supply, reducing its volume and mass. It uses analogue-to-digital current sensors to provide the user with information on power usage by the individual components. This will allow a trade-off system whereby users can decide if they want to concentrate power in specific areas by reducing usage in others. The design is modular so that the two card slots can be swapped to exchange programmes or to use external facilities, such as a digital camera.

There is a tendency in designing human-centred machines for designers to make certain assumptions about the consumer. Almost invariably they underestimate people's ability to use the technology. IBM has a Usability Research Lab at its Almaden Research Centre in San Jose. There, engineers are required to stand behind glass mirrors and watch ordinary people try to use their software. The department head, Dan Russell, describes it as 'torture for the engineers.' Cartoonist Scott Adams in *The Dilbert Future*, 1997, describes the frustration of the user:
'When I was a kid, I had a little black-and-white Sears television set that was very easy to use. There were only three steps:
1   Turn on power.
2   Select channel.
3   Wrap a long string to the horizontal hold knob on the side of the television so I could continually adjust it from across the room by pulling the string with my feet.
Those simple days are gone. Now I have a home entertainment center. It has six remote controls. If I want to watch television using the satellite dish as source, I have the following steps:
1   Hire Sherpa guides...'
Adams' description continues, but the story will be widely echoed by other rueful users.

This new interactive technology is not only forcing new approaches to design, but also requires non-designers to contribute different

skills. Beatriz Ayala has conducted research to look at the way in which elderly people interacted with new technology in cars. The assumption was that people who had never used computers during their working lives would have difficulty operating Information Technology (IT), such as in-car navigation aids. Although one woman immediately reversed her car into a bollard thinking that a navigation aid would park the car for her, the findings of the survey were that elderly people were quite comfortable using IT systems and adapted to them very quickly. The Ford Motor Company is now wary of making the wrong assumptions, and has developed a special suit to help its (young) designers design cars for older people. The Third Age Suit simulates the experience of being seventy years old with restricted joint movement and surgical-style gloves to mask tactile sensation. Vision is reduced by visual stimulation spectacles tinted yellow, increasing the wearer's sensitivity to glare. The company has already produced one car, the Ford Focus, using the suit.

In the developed world the population is ageing as the birth rate decreases and life expectancy increases. The healthcare industry is growing dramatically as a result, coupled with growth in care for the elderly. Traditional family practices, whereby the elderly were cared for by younger women relatives who did not go out to work, are disappearing, and it is increasingly the state that is having to provide carers.

14

15

14••
The HelpMate robot, designed by Joseph Engelberger, founder of modern industrial robotics, is produced by Pyxis Corporation. Its sensors allow it to move without tracks safely around a hospital with medical supplies, food and information for staff and patients.

15••
Tactile display systems may provide a more user-friendly and ergonomic alternative to traditional designs. These work by stimulating the skin's perceptual nerves. In 1999/2000, at Carnegie Mellon University, Francine Gemperle, Nathan Ota and Dan Siewiorek developed a navigation aid that gives the simple commands forward, back, left, right.

16••

Not all products focus on the young
consumer. Ford have developed
a Third Age Suit to give its
designers some idea of what it
is like to be an elderly driver.
The suit restricts joint movement,
gloves muffle tactile sensation,
and spectacles tinted yellow
reduce vision and increase
the wearer's sensitivity to glare.
One car model, the Ford Focus,
designed with the aid of the suit
has already been put into
production.

17 + 18●●

In 1998 fashion designer Alexander McQueen and photographer Nick Knight were responsible for one of the most groundbreaking issues of British magazine *Dazed and Confused* for which McQueen was guest editor. Physically disabled models were the main feature, and double amputee Aimee Mullins appeared on the front cover wearing van Philips's C-shaped prostheses. Mullins went on to appear in McQueen's Spring/Summer 1999 catwalk show, wearing custom-build prostheses especially designed for her by McQueen. Giving her reasons for wanting to appear in the magazine and on the catwalk, Mullins commented, 'I don't want people to think I'm beautiful in spite of my disability but because of it. It's my mission to challenge people's concept of what is and isn't beautiful.'

Robot carers in the home are seen as a way of providing an alternative to institutional care. This is leading to the development of robots equipped with an Artificial Intelligence that enables them to provide both companionship and medical support through health monitoring. A primary consideration in the design of these robots is acceptance by the elderly consumer.

Scientists at Carnegie Mellon University and the University of Pittsburgh in Oakland are developing such a robot, called Nursebot. It is designed to monitor medication, and to call for human aid if needed via an electronic link using the internet and video conferencing. This robot is equipped with two cameras disguised as eyes, while motion and location will be monitored by an invisible light broadcast from the base of the machine. The beam works like sonar bouncing against legs and feet of the residents. Much of this current research is intended to determine whether or not the robot will be accepted by the elderly. Sebastian Thrun, assistant professor of computer science at Carnegie Mellon University recently admitted: 'Replacing human contact [with a machine] is an awful idea. But some people have no contact [with caregivers] at all. If the choice is going to a nursing home or staying at home with a robot, we think people will choose the robot.'

It is not always desirable to have a product designed to fit in with people's preconceptions. Some of the most innovative designs actually go against existing wisdom. Van Philips, director of the prosthetic design company, Flex-Foot, is himself an amputee. Among his designs is the Sprint-Flex III lower leg prosthesis for use in athletics. The design is driven by comfort and performance. Yet it makes no attempt to mimic the appearance or movement of a human limb. Instead, the designer studied the fastest animal on Earth, the cheetah. Interviewed in *ID* magazine (May 1998) Van Philips describes the cheetah as it runs: 'The animal lands on the front and transfers weight to the rear, the C-shape closes up and collapses, stretching those tendons on the backside of the leg.' Sprint-Flex mimics this in its C-shape design. The carbon-fibre prosthesis gives the athlete a spring-like movement, allowing him or her to move much faster than with a regular prosthesis.

Robot designs tend to be either crude metal boxes, or humanoid in behaviour or appearance. The former are omnipresent in factories, while the latter are being developed as companions with the idea that people will feel most comfortable with something that looks familiar. At a workshop on smart materials held at the Netherlands Design Institute, the robotics group set about finding an alternative to both of these. A multidisciplinary design team was headed by service robot expert K. G. Engelhardt. Acknowledging that one of the problems with service robots is user acceptance, the team designed a robot with all the usual behaviour and function characteristics, but with a more user-friendly appearance. The group looked more to the design of robot pets, such as the Tamagotchi, for inspiration than the 'black box' or humanoid design approach. The result was a Parrot Scarf, which like the pirates' parrots would look out for the welfare of their owners, alerting them to changes in their environment or providing them with information useful for their activity. A modular approach to design meant that the robot could be supplied as a basic unit, with the users adding functions to customize the scarf according to their particular needs.

# new ways
# of working

Alongside the cyborg-related technologies, market demands have given rise to the development of many new scientific partnerships. This can be seen as part of a growing trend towards technologies that are multidisciplinary. Many of these technologies are not significant when looked at in isolation, but become of critical importance when coupled with other technologies. One of the most striking examples is linking the computer and responsive technologies to get a combination that is effectively a smart system. Responsive or 'smart' materials have the ability to sense and respond to their environment. These can include sensors and actuators, which in turn are linked by a computer that processes the information, allowing it to flow between them. Until recently, many smart materials and systems were dismissed as 'answers looking for questions', doomed to gimmick products such as T-shirts that change colour when touched.

Military and space-related industries have traditionally funded much of the research and development in cyborg technologies. The end of the Cold War, coupled with greater accountability in public spending in the West, have changed this. There is growing pressure to find commercial applications for many of these developments, partly in order to try and to recoup some of the money invested, but also as an exercise in public relations. Whatever

the politics, these developments should bring great benefits to many sectors of society.

NASA was one of the first to disseminate its technologies for the benefit of commercial and public interests, and has operated a Commercial Technology Program for over thirty years. The organization annually publishes *Spinoff* to highlight NASA technology that has been successfully commercialized. NASA's technologies have found applications in consumer products, industry, computer technology, the environment, as well as health and medicine. From NASA's point of view, demonstrating that the same technology that helps an astronaut adapt to conditions in space could also help cancer patients is a good way of illustrating the relevance of the space budget allocation to the American public, and in addition, commercial industry, and small businesses in particular, have access to leading technologies that they would not have the budget to research and develop themselves.

Enabled by a NASA Small Business Innovation Research (SBIR) contract from the Jet Propulsion Laboratory, a robotic arm that can assist surgeons in non-invasive endoscopic procedures has been developed by Computer Motion Inc. The Automated Endoscopic System for Optimal Positioning, or AESOP, allows surgeons to control the motion of the camera, which is attached to a robotic arm.

Voice-recognition software is pre-recorded on to a voice card and inserted into the controller. This follows research showing that voice-controlled commands are preferred in the operating room to eye-tracking and head-tracking (controlling motion in response to movements of the surgeon's head). NASA hopes to make use of the technology powering the robotic arm of AESOP to service satellites and inspect payloads on the Space Shuttle in the future. The aim is to use robotics on space repair missions requiring exact and precise movements that exceed human dexterity.

An SBIR grant has helped another company, Delsys Inc of Boston, Massachusetts, to develop a MyoMonitor electromyography (EMG) system that can be used to monitor muscular activity when the person is moving. For NASA the negative effect of microgravity on muscle tissue is a constant problem for astronauts. Sports medicine and exercise training also benefit from the use of a mobile EMG system which allows athletes to be assessed when in motion rather than having to stand still. Applications can also be envisaged, an example being helping to prevent work injuries, such as repetitive strain injury. The MyoMonitor uses an active parallel bar electrode which does not need any skin preparation or conductive gel. It is also more robust than traditional EMG equipment and is unaffected by vigorous movement of the body or perspiration.

19 + 20●●
One company to benefit from NASA's Spinoff programme is California-based Computer Motion Inc. The company has developed an endoscopic procedure based on robotics which is designed to be less invasive for the patient and lead to a faster recovery time. AESOP, or given its full title, the Automated Endoscopic System for Optimal Positioning uses a robot arm to allow surgeons to control the motion of the endoscopic camera. The surgeon can communicate with the robot using voice-recognition software. The use of the robot gives the surgeon a steadier view of the surgery and more precise camera movement than would be possible if it were hand-held. The smaller image shows the next generation of this technology in the ZEUS Robot Surgical System. Additional robot arms allow the surgeon to perform the procedure with even greater precision, making incisions smaller than the diameter of a pencil.

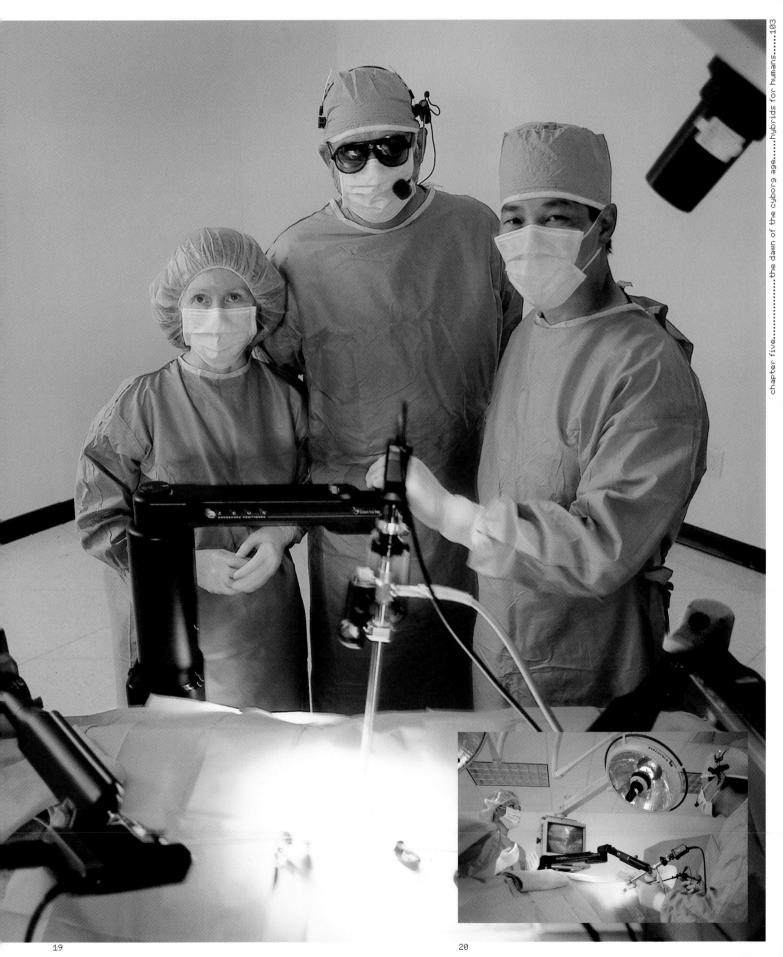

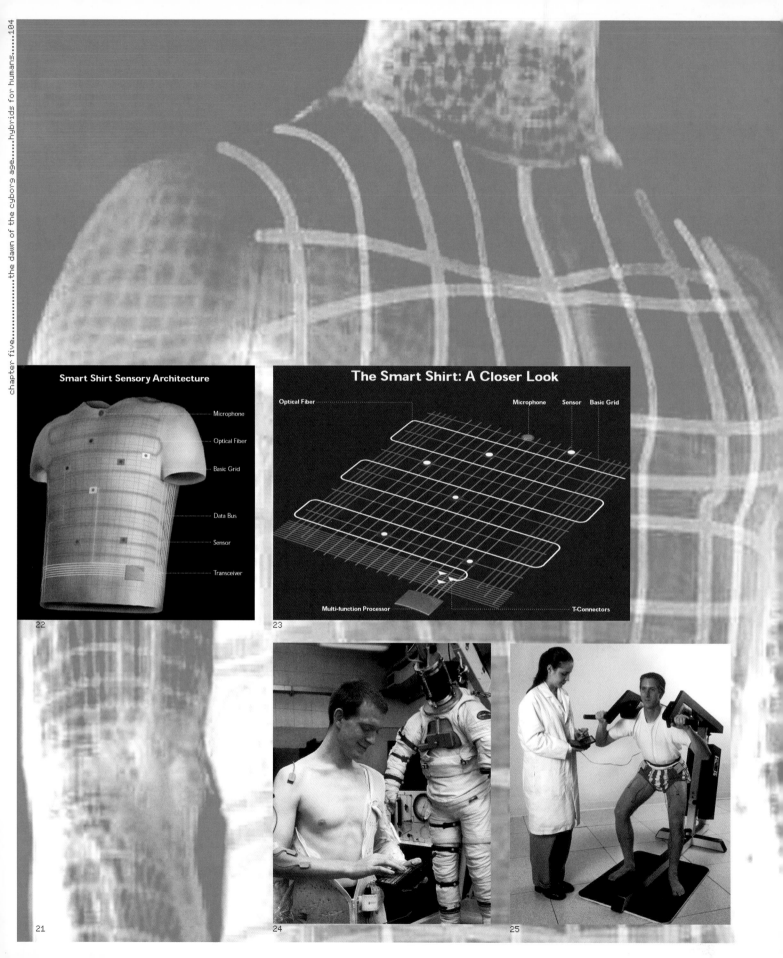

**Smart Shirt Sensory Architecture**

Microphone

Optical Fiber

Basic Grid

Data Bus

Sensor

Transceiver

22

**The Smart Shirt: A Closer Look**

Optical Fiber

Microphone     Sensor     Basic Grid

Multi-function Processor

T-Connectors

23

21

24

25

Once the information is gathered it is then transferred to a computer for analysis.

For security reasons, dissemination of materials and technologies from military sources is more problematic than from space exploration . The US Defense Advanced Research Projects Agency (DARPA) funded initial research into the development of a 'smart' T-shirt. The intention was to develop a health-monitoring garment that marines could wear in combat. In the case of injury a 'sensate liner' would relay information about the injury and the location of the marine to a remote medical triage unit (which assesses treatment priority). The T-shirt was intended to improve efficiency and save lives by informing medical personnel in advance whether an injury affected a vein or artery and how close it was to the heart. The Smart Shirt incorporates a 'Wearable Motherboard' which is essentially a fabric embedded with sensors that can receive and transmit information between the wearer and garment. This is used alongside an especially developed Application Programming Interface (API), open architecture hardware and software, which then relays information from the garment to a remote (wireless) location such as the medical triage unit. Apart from reliability, other concerns in developing the garment were robustness, weight and comfort.

A company has now been set up to bring the Smart Shirt technology to non-military applications. Sensatex Inc, New York, includes the Georgia Tech Research Corporation as equity stakeholders in recognition of the university's contribution in researching and developing the technology. The company has identified a number of market areas for development. Each has slightly different requirements. A Child Monitor would operate as a digital umbilical cord, allowing parents to keep track of their child's location through a two-way voice communication system. The proposed Sensate Geriatric Monitor is another reassurance product designed for elderly people. Here a continuous voice-communication system is less appropriate, and sensors that can monitor vital signs and location are more important for dealing with any accidents or emergency situations. Other applications envisaged by the company include Personal Area Networking (linked wearable technology devices located around the body), and cardiovascular and hazardous occupations monitoring. Each would operate on the same basic principle, but adapted for specific needs in every application.

21 + 22 + 23••
The technology for Smart Shirt was originally undertaken for military applications, intended to help save lives by informing medical personnel of a soldier's injury and location. The fabric includes a complex grid of fibre-optic sensors that can be enhanced with further monitoring devices specific to each application and the information needed. Sensatex Inc are now looking to developing commercial products, and have identified possibilities in the marketplace for childcare and the care of the elderly.

24 + 25••
Some of the problems experienced by astronauts are similar to those faced by patients recovering from illnesses. The monitoring of muscles is vitally important to the astronaut , who must exercise regularly while in space to prevent muscle shrinkage. A similar monitoring system can also be used to assist in rehabilitative therapy, prevention of injury in the workplace and sports medicine. Delsys Inc have developed a MyoMonitor electromyographic (EMG) system that offers a wearable four-channel device that is smaller and less cumbersome than static machines.

26••
Using the body as a potential source
of power or energy is not entirely
new. The Whirling Dervishes date
from the thirteenth century, and
their beliefs and rituals continue in
Turkey today largely unchanged, as
shown here. It is believed that when
the dervish is in prayer his body is
open to receive the energy of God.
The term dervish means 'doorway',
and the dervish's role is to act as
as a conduit for the power that he
merely transmits to earth.

27••
One issue with wearable computers
is the amount of power needed to
operate them. This early Wearcom
wearable computer from Steve Mann
has its power supply and operating
system (a Pentium II processor) stored
in the backpack. It is shown as
exhibited in the exhibition at
the Stedelijk Museum, Amsterdam, '
'The Soft Machine: Design in the
Cyborg Age', 1998/9.

28••
Visually enticing as it may seem,
our future is unlikely to resemble
Stanley Kubrick's film 2001. The
reality of the year 2001 was that
voice-recognition software was still
a rarity, and Artificial Intelligence
of HAL's sophistication is still a
dream. By 2001, of all the futuristic
aspects of the film, it was only Arne
Jacobsen's stainless steel cutlery
that had become a design classic.

29••
The future of clothing is likely to
combine performance, aesthetics
and medicines. In this concept for
a fabric that releases vitamins or
medicines on to the wearer's skin
for absorption, British designer
Lesley Dennyson has grown crystals
on a synthetic fabric. The result is
a design where the performance
becomes the aesthetic.

# human-powered machines

The question of how to power wearable technology has yet to be solved satisfactorily. Batteries are becoming increasingly powerful, but recharging remains an issue, particularly for global travellers who have to carry adaptors for different power supplies. Among the new methods under consideration is human power that would be generated by the wearer.

The idea of using a 'body network' to communicate without wires was put forward by Thomas G. Zimmerman in an article written for the IBM Systems Journal in 1996. His idea was that people could generate their own power supply, and even contribute any excess created to a type of 'national grid' so that none would be wasted. Others have followed on these thoughts, including Thad Starner. He has looked at the possibility of generating power from breathing, body heat, blood circulation, arm motion, typing and walking. The only method that he considers might be viable is the last of these – walking. Starner concluded: 'Using the legs is one of the most energy-consuming activities the human body performs. In fact, a 68 kg man walking at 3.5 mph, or 2 steps per second, uses 280 kcal/hr or 324 W of power.' The problem remains of how to harness and store any energy created.

Typically, new technologies are introduced as a replacement for an existing product. But it can take some time before such technologies really establish their own identity and value. Parallels can be drawn between the development of wearable technology that we are now seeing and the introduction of synthetic materials in the early part of the twentieth century. Plastic, for instance, was first promoted as a low cost alternative to ivory, and consequently became associated with cheap products. It was not until the end of the last century that designers took the material seriously and designed upmarket goods that took advantage of the finer qualities of plastic.

The next few years will be crucial in determining whether wearable technologies will simply be another way of tethering humans to the internet – extensions of existing products such as the telephone, computer and television. Alternatively, wearables may well be the portal through which the broader spectrum of cyborg advances can rapidly be established in the marketplace.

Research and development thus far has been primarily driven by military, space, medical and electronic consumer product interests. It is through the last that the technology is being most widely disseminated and made generally available. The first wave of products to reach the market were largely extensions of existing products. The hybrid watch/telephone/washing machine (well almost) were not user-friendly or ergonomic, nor were they stunning pieces of design. Rather, they reflected the designer/technologist's penchant for science fiction films, whose references gave an unmistakable retro-futuristic look from somewhere in the middle of the twentieth century. Thankfully, this early aberration seems to have passed, and we are now seeing more thoughtfully designed products reaching the market. These are proving very seductive indeed, and draw aesthetics and technology together to create exciting new design.

It must be remembered that these products are primarily intended to enhance our relationship with the world around us and with one another. As such they create an interface that must not come to dominate. This is not a design or even a technology issue, but rests with the users, the same users who cannot be persuaded to install their software correctly, and who in a restaurant choose to speak on their mobile phone rather than to friends at the same table. We do not experience elements in the world around us in isolation – everything is inter-linked. As technology strengthens the bond between us and our environment, it, too, becomes part of our environment, and by extension, part of what we are about to become.

We are now in a position to decide if and how we want to embrace these technologies. With power comes a collective responsibility both to ourselves and to the environment we live in. Utopia or dystopia, the choice is ours••

# further reading

## books

Beckmann, John, *The Virtual Dimension*, New York, 1998

Bester, Alfred, *Tiger! Tiger!*, Harmondsworth, London, 1956

Bulgakov, Mikhail, transl. from the Russian by Michael Glenny, *The Heart of a Dog*, London & New York, 1968

Burgess, Anthony, *Clockwork Orange*, London & New York, 1962

Braddock, Sarah E., and Marie O'Mahony, *TechnoTextiles: Revolutionary Fabrics for Fashion and Design*, London, 1998

Brouwer, Joke, and Carla Hoekendijk, *Technomorphica*, Rotterdam, 1997

Darwin, Charles, *The Origin of the Species by means of Natural Selection*, London, 1859

Derycke, Luc, and Sandra Van de Veire (eds), *Belgian Fashion Design*, Ghent, 1999

Diller, Elizabeth, and Ricardo Scofido, *Flesh*, New York, 1994

Dunne, Anthony, *Hertzian Tales: Electronic Products, Aesthetic Experience and Critical Design*, London, 1999

Eliade, Mircea, transl. from the French by Willard R. Trask, *Shamanism: Archaic Techniques of Ecstasy*, revised and enlarged edn, London & New York, 1964

Ettinger, Robert C. W., *The Prospect of Immortality*, London, 1965

Ewing, William A., *Inside Information: Imaging the Human Body*, London & New York, 1996

Fuller, Errol, *Extinct Birds*, Oxford, 2000

Gabriel, Richard, *No More Heroes: Madness and Psychiatry in War*, New York, 1987

Grant, P. R., *Ecology and Evolution of Darwin's Finches*. Princeton, NJ, 1986

Graves, Robert, *The Greek Myths*, 2 vols, revised edn, Harmondsworth, 1962

Haraway, Donna J., *Simians, Cyborgs, and Women: The Reinvention of Nature*, London & New York, 1991

Harris, Philip Robert, *Living and Working in Space: Human Behavior, Culture, and Organisation*, 2nd edn, Chichester, Wiley, 1996

Hoffmann, Dr Heinrich, *Struwwelpeter*, (first published in German 1844, and in English 1848) English edn, London, 1995

Kafka, Franz, transl. from the German by Willa and Edwin Muir, *Metamorphosis and Other Stories*, Harmondsworth 1961; London 1992

Levidow, Les, and Kevin Robins (eds), *Cyborg Worlds: The Military Information Society*, London, 1989

Menzel, Peter, and Faith D'Aluisio, *Robo Sapiens: Evolution of a New Species*, Cambridge, Mass., 2000

Neumann, Dietrich (ed.), *Film Architecture: From Metropolis to Blade Runner*, Munich, 1996; London & New York, 1999

Orta, Lucy, *The Process of Transformation*, Paris, 1995

Osterwalder, Anja, *Space Manual*, Stuttgart, 1999

Ovid, transl. from the Latin by A. D. Melville, *Metamorphoses*, Oxford, 1987

Phillips, Gordon, *Best Foot Forward*, Cambridge, 1990

Picard, Rosalind, *Affective Computing*, Cambridge, Mass., 1997

Piercy, Marge, *Woman on the Edge of Time*, New York, 1976; London, 1979

Redhead, David, *Products of Our Time*, Basel & London, 2000

Sanderson, Peter, *Marvel Universe*, London & New York, 1996

Simons, Geoff, *Robots: the Quest for Living Machines*, London & New York, 1992

Stephenson, Neal, *The Diamond Age*, New York, 1995; London, 1996

Stevenson, Robert Louis, *The Strange Case of Dr Jekyll and Mr Hyde*, first published London, 1886

Stork, David E., (ed.), *Hal's Legacy: 2001's Computer as Dream and Reality*, Cambridge, Mass., 1997

Turney, Jon, *Frankenstein's Footsteps: Science, Genetics and Popular Culture*, London & New Haven, Conn., 1998

Virilio, Paul, *Lost Dimension*, New York, 1991

Warner, Marina, *No Go The Bogeyman*, London, 1998; New York, 1999

Warner, Marina, *Managing Monsters: Six Myths of Our Time*, London, 1994; New York, 1995

Wells, H.G., *The Short Stories of H. G. Wells*, London, 1927

## exhibition catalogues

*Robots*, Foundation Cartier pour l'art contemporain, 30 June–14 November, 1999

*Soft Machine: Design in the Cyborg Age*, Stedelijk Museum of Modern Art, 28 November 1998– 10 January 1999

*The Art of Detection: Surveillance in Society*, MIT List Visual Arts Center, 9 October– 28 December, 1997

## conference + academic papers

IEEE Computer Society, *The First International Symposium on Wearable Computers*, 13–14 October 1997, Cambridge, Mass.

# photo credits

pp. 4–5 Most of the images on these pages appear elsewhere in the book, and credits are under those page references. The exception is the Walter van Beirendonck image, third from the right, from the series *Starship Party*, taken by photographer Robert Stoops;
p. 15: James Hyman Fine Art, London (06);
p. 19: Tom Barker (02); p. 20: Wellcome Institute Library (03), John Farnham (04), Prana-Film Gmbh, Berlin (courtesy Kobal) (05);
p. 21: Tomb of Thoutmosis from *The Royal Mummies*, by G. Elliot Smith, 1912 (07);
p. 22: Mark L. Sabo (09); p. 23: courtesy J. B. Bacquart (11); p. 26: Universal (courtesy Kobal) (15), Victor Habbick Visions/Science Photo Library (SPL) (16); p. 27: Warner Brothers (courtesy Kobal) (17); p. 28: 20th Century Fox (courtesy Kobal) (18); p. 30: Kevin Warwick (19); pp. 32–33: John Kaine;
p. 34: John Farnham (02), drawing from *The Maya* by Michael D. Coe published by Thames & Hudson Ltd (03); p. 35: John Farnham (04); p. 36: Wellcome Institute Library (06); p. 37: Universal (courtesy Kobal) (08); 20th Century Fox (courtesy Kobal) (09);
p. 38: Science Photo Library (10); p. 39: BAE Systems (11); p. 41: Lucasfilm/20th Century Fox (courtesy Kobal) (13), Brian Brake/Science Photo Library (14), MGM (courtesy Kobal) (15);
p. 42: Peter Menzel/Science Photo Library (16, 17 + 18); p. 42: courtesy Laurie Anderson (19);
p. 43: Ladd Co/Warner Brothers (courtesy Kobal) (20); p. 44: Honda UK (21), Stelarc (22);
p. 45 Sony (23); p. 46: Wellcome Institute Library (24); p. 47: Ronald Stoops (25);
p. 48: Ronald Stoops (26); p. 49: Nick Knight for Comme des Garçons (28); p. 52: Dr Andrew Forge/Wellcome Photo Library (01);
p. 53: Paramount (courtesy Kobal) (02), Universal (courtesy Kobal) (03);
p. 55: Blatchford (05), University of Washington (06), Peter Kyberd (07); pp. 56 + 57: Alexa Wright (09 + 10); p. 58: Medtronic (12);
p. 59: University of Washington (13 + 14);
pp. 60 + 61: European Space Agency (ESA) (15 + 16); p. 62: Lucy Orta; p. 63: Jacqueline Alos (18); p. 64: ESA (19), Paramount (courtesy Kobal) (20); p. 65: ESA (22); p. 66: Steve Mann (23), Niels Bonde (24), Emma and Jane Hauldren (25); pp. 68 + 69: Motorola (26);
p. 73: Universal (courtesy Kobal) (01);
p. 74: Simon Fraser/Royal Victoria Infirmary,

Newcastle upon Tyne/Wellcome Institute Library (02); p. 77: Producers Associates (courtesy Kobal) (03), Wellcome Institute Library (04); p. 79: United Artists (courtesy Kobal) (05); Kaiju Theatre Production (courtesy Kobal) (06); p. 80: Juergen Berger, Max-Planck Institute/Science Photo Library (07), John Farnham (08); p. 83: James King-Holmes/ICRF/Science Photo Library (09);
p. 86: Wellcome Institute Library (11), Dan Lecca (12 + 13); p. 87: Wellcome Institute Library (14); p. 88: Photo W. Suschitzsky (15); Eden Ostrich World (16); Rumpus (18);
pp. 90–91: UFA (courtesy Kobal); p. 92: Dunn + Raby (01); Lubna Hammond (02); p. 93: Tom Barker (03), Simon Costin (04); p. 95: Warner Brothers (courtesy Kobal) (08), UFA (courtesy Kobal) (09); p. 96: Carnegie Mellon University (11); p. 97: Steve Mann (12); p. 98: Pyxis Corporation (14), Carnegie Mellon University (15); p. 99: Ford Motor Company Ltd (16);
p. 101: Photo Nick Knight and Art Direction Alexander McQueen for *Dazed and Confused* (17 + 18); p. 103: Computer Motion Inc (19 + 20); p. 104: Sensatex (21–23); Kevin Wilson (24 + 25); p. 107: Turkish Tourist Board (26), MGM (courtesy Kobal) (28), Lesley Dennyson (29).

Every effort has been made to identify the copyright holders of the illustrations included in this book. The publishers would be grateful to be informed of any omissions.

Page references of illustrations are in italics, followed by the picture numbers in brackets

Thames & Hudson
£18.95